MW00574847

Make:
More Tinkering

Make:
More Tinkering
Kids in the Tropics Learn by Making Stuff

Curt Gabrielson

Foreword by Brian Schmidt, Nobel Laureate for Physics

Maker Media, Inc.
San Francisco

Published by
Maker Media, Inc.
1700 Montgomery Street, Suite 240
San Francisco, CA 94111

Maker Media books may be purchased for educational, business, or sales promotional use. Online editions are also available for most titles (safaribooksonline.com). For more information, contact our corporate/institutional sales department: 800-998-9938 or corporate@oreilly.com.

Editorial Director: Roger Stewart
Copy Editor: Elizabeth Welch
Proofreader: Scout Festa
Interior and Cover Designer: Maureen Forys,
    Happenstance Type-O-Rama
Illustrator: Richard Sheppard, Happenstance
    Type-O-Rama
Indexer: Valerie Perry, Happenstance Type-O-Rama

October 2018: First Edition

Revision History for the First Edition

2018-10-15 First Release

See oreilly.com/catalog/errata.csp?isbn=9781680454369 for release details.

978-1-68045-436-9

Safari® Books Online

Safari Books Online is an on-demand digital library that delivers expert content in both book and video form from the world's leading authors in technology and business. Technology professionals, software developers, web designers, and business and creative professionals use Safari Books Online as their primary resource for research, problem solving, learning, and certification training. Safari Books Online offers a range of plans and pricing for enterprise, government, education, and individuals. Members have access to thousands of books, training videos, and prepublication manuscripts in one fully searchable database from publishers like O'Reilly Media, Prentice Hall Professional, Addison-Wesley Professional, Microsoft Press, Sams, Que, Peachpit Press, Focal Press, Cisco Press, John Wiley & Sons, Syngress, Morgan Kaufmann, IBM Redbooks, Packt, Adobe Press, FT Press, Apress, Manning, New Riders, McGraw-Hill, Jones & Bartlett, Course Technology, and hundreds more. For more information about Safari Books Online, please visit us online.

How to Contact Us

Please address comments and questions to the publisher:

Maker Media, Inc.

1700 Montgomery Street, Suite 240

San Francisco, CA 94111

You can send comments and questions to us by email at books@makermedia.com.

Maker Media unites, inspires, informs, and entertains a growing community of resourceful people who undertake amazing projects in their backyards, basements, and garages. Maker Media celebrates your right to tweak, hack, and bend any Technology to your will. The Maker Media audience continues to be a growing culture and community that believes in bettering ourselves, our environment, our educational system—our entire world. This is much more than an audience, it's a worldwide movement that Maker Media is leading. We call it the Maker Movement.

To learn more about Make: visit us at make.co.

This book is dedicated to the immortal spirit of Paul Doherty (1948-2017), who encouraged me to explore scientifically to the ends of the earth, and to always, always pass it on.

ACKNOWLEDGMENTS

Thanks × 10^6 to those who made this book better: Stephen Thompson, the jolly Irish geologist who taught me about Timor's fascinating rocks. Julie Yu for sundry lucid chemistry clarifications. Karen Kalumuk for similar biological explanations. Bill Maney for checking my math and physics. David Keith for climate pointers. Roger Stewart for patiently guiding it down the road to publication. Elizabeth Welch, copy editor extraordinaire. Scout Festa, the supernatural proofreader. Maureen Forys, master of the layout.

The SESIM crew here in Dili, Timor-Leste: Luis, Vero, Mimi, Caetano, Bernardino, Angela, Olandino, Julio, and many other inspirational Timorese teachers. All my colleagues at the Timor-Leste National Commission for UNESCO, under the Timor-Leste Ministry of Education.

Frances Gabrielson for giving input, albeit far outside her considerable realm of expertise. My partner Pam for ever-cogent editing, as well as pulling more weight in the household as I played with this stuff.

My fabulous tropical models, friends of my kids:

ACEDE   ELIS   MARIÇA   NOVA   TANIA   WILTOR

And of course my own kids!

PAULO RAMELAU   ZORAYA ALICIA

CONTENTS

FOREWORD

Timor-Leste is one of the world's youngest democracies. Having emerged out of years of conflict, it is a nation poor when measured in GDP, but rich in human resources. My fellow Nobel Prize winner, the remarkable José Ramos-Horta, invited me to Timor-Leste in 2013, when I got my first taste of the country.

As part of that visit, I met Curt Gabrielson for the first time. He translated my talk about cosmology to a few hundred teachers, who were assembled from across the country to learn about teaching science. His translation, in Tetun, connected humanity's understanding of the universe to many of the nation's science teachers. Some had felt this feat would not be possible in Tetun, but it was done to great effect, as we were able to see from the reactions of the teachers.

I returned to Timor-Leste in 2017 for Ramos-Horta's Nobel Prize celebration—one of several laureates this time. On this trip, I witnessed the diligence and perseverance of the Timorese to learn whatever is necessary to make their nation grow and develop despite the incredible obstacles and hardships they've faced. Great progress over those four years gave me great confidence in the future.

On this trip, Curt arranged to have me speak a few words on record to support the new curriculum the Ministry of Education had created for primary and junior high schools. This is the first truly indigenous curriculum Timor-Leste has had, and it is full of hands-on lessons that link the concepts being taught to the daily life and experiences of the Timorese. It is exactly what is required to help teach the young generation the skills they will need to prosper in the future.

This book of activities from Curt and his group of science and mathematics teachers in Timor-Leste draws primarily from the topics developed for that new curriculum. It is a wonderful opportunity to learn more hands-on science and to learn more about the world around us.

I'm happy to see Curt using the rich science and mathematics of Timor-Leste to increase the quality and effectiveness of schooling there. In this way, the Timorese can be a model for the entire global education community. Well done, Curt! And I look forward to our next visit.

BRIAN P. SCHMIDT
2011 Nobel Laureate in Physics
Vice Chancellor and President,
Australian National University

INTRODUCTION

Tinkering is one of my favorite things to do. I grew up tinkering on a hog farm in Missouri. I got into a decent college and tinkered my way through a physics degree. Since then I've tinkered in several countries; for hire and pro bono; with street urchins and PhDs; below sea-level and at 5,000 meters; within elite institutions and with ragtag gangsters; toward lofty goals of enlightenment as well as cheap thrills; with close friends and total strangers; with high-tech paraphernalia and also art, music, and plumbing supplies; at all times day and night; and in various states of dress and undress. It's all been magnificent and I highly recommend it.

So I wrote a book about my experiences tinkering, especially what I learned running a Community Science Workshop in the rural immigrant community of Watsonville, California. Called *Tinkering*, it's full of projects and also info on how to run a small science center.

During and after my time in Watsonville, I was involved in pumping up the science and mathematics education in the tiny new nation of Timor-Leste (East Timor). There the National University and the Ministry of Education both asked me to help develop curriculum and train their science and mathematics teachers, so I have been tinkering with that on and off since 2000. I've lived here 10 years over the last 17, and steady since 2012 with my family. In my time in Timor I have learned a heck of a lot more about

tinkering, which I'll do my best to pass on to you in this book.

A lot of the tinkering I've done here has been in conjunction with a remarkable group of science and mathematics teachers known as SESIM. The acronym stands for the Center for the Study of Science and Mathematics, a curiously pompous name for these tireless young teachers pushing toward a better way to do education. We call what we do *pratika*. This is a Tetun word from Portuguese that has come to mean anything hands-on, experiential, lab centered, inquiry based—essentially learning through exploring and investigating; that is, tinkering.

In the United States, there are mobs of people working on doing this sort of education, each with a special tactic, some proclaiming their way to be the one true path to enlightenment. In Timor, essentially nobody else is doing education through tinkering. It's all chalk and talk in the classrooms, so anything *pratika* is positive. In a decade or so when everyone here is using real stuff in the science and math classrooms, and kids are all comfortable learning from their own observations of experiments, and teachers are happy receiving and responding to a torrent of questions from their students, we'll sort out the finer pedagogical distinctions between the 15 or so different inquiry approaches. For now we just play and learn with real stuff and have a rollicking good time at it.

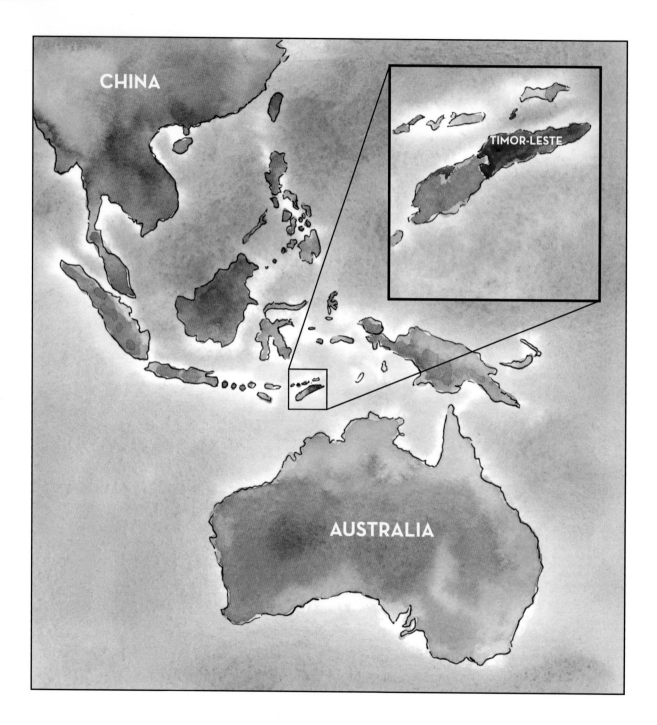

I've worked together with the SESIM teacher-trainers to develop hundreds of *pratika* teaching activities, given dozens of seminars and workshops, and written some 20 manuals full of *pratika* activities (in Tetum, the lingua franca of Timor-Leste). We recently got ourselves a genuine lab at a large public school in the capital and filled it with interesting junk for tinkering. Local teachers bring their students there, and SESIM uses it as a national science center to support teachers in other districts working to create their own small science centers. I've documented some of our work here: *www.timorpratika .wordpress.com.*

One of the most fascinating parts of this experience for me was learning to use Timor's set of cheap and free stuff to tinker. In the United States, I've long been accustomed to dumpster diving, gleaning from the streets and ditches, and keeping a keen eye on people's trash.

Here in Timor, trash heaps do not hold nearly the same potential as dumpsters in the United States. On the whole, people tend to use stuff longer, fix it again and again, and throw things out only when they're really trash. Then a crew of hard-up people roam the streets, making their living off what few valuable things do remain in the trash heaps.

So my colleagues and I have developed activities based on much more meager trash, and we've found other sources of cheap and free materials. One such source is the jungle; we've got *pratika* employing bamboo, banana leaves and branches, various parts of various palm trees, coconuts, rock-hard sago nuts, and the earth itself—sand, clay, stones, and pebbles of all sorts. Worry not: I'll mention substitutions if you have trouble finding some of these in your temperate habitat, but the point is this: if you can't find the ideal material, get creative!

Don't spend much money; it's not sustainable, and sustainability is key. It's one of the three foundational principles of SESIM that make education work well. The other two are relevance and concreteness. Relevance should be pretty obvious, though it remains a real challenge when using imported textbooks. Concrete concepts are not some sort of limitation but rather the starting point; you should absolutely make the leap to the abstract eventually.

Warning: I've written some of these activities to a conceptual depth not common in how-to books; feel free to skim over the science explanations if you'd rather get on with more tinkerings. A few of these activities are not traditional tinkering topics. Free your mind! Food calendars, basketry math, and slingshot physics are absolutely tinkerable, and I'd argue they hold even more value than a 3D-printed Arduino-controlled blinky.

The activities here were developed by SESIM and myself over the last few years, and include influence from far and wide. Feel free to nab any one of them and run to the distant corners of the world.

SAFETY

Various tinkerings in this book can hurt you if you're not careful—maybe even seriously. Safety is always first in any successful tinkering session, and here are a few areas of special concern as you use your level head to always be aware of risky situations:

▸ **Hot glue hurts when it touches your skin, and the tip of the glue gun does, too.** It will give you a big white blister and take the whole top layer of skin off. This can be quite a learning experience, but most agree that it's not worth the pain. Using low-temperature hot glue will always work for the activities in this book, and small glue guns are easier to control than large ones.

▸ **Electricity can kill you.** None of the tinkerings in this book will connect directly to outlet voltage, but even if you're just using the hot-glue gun, you should be mindful that the wires inside the insulation have the potential to stop your heart if you happen to touch them.

▸ **Open flames can burn your house down.** Don't let them. Even if it's just a candle, always have an adult on hand before lighting up. Never experiment with a candle or other flame on surfaces of wood, plastic, or cloth, and never in an environment where there may be other gases or liquids that may catch fire or explode. Having a fire extinguisher on hand is always a good idea. It saved my house once—see Chapter 2.

▶ **Various tools and materials indicated in this book will cut, hack, pinch, poke, impale, or gash your tender skin.** Don't let them. You're in charge, boss, so it's up to you to make the tool do what it's supposed to without hurting you.

▶ **Some operations described in this book with high pressure or tension can launch projectiles, big and small, which can hurt you seriously.** Avoid that situation. Your eyes are especially valuable as body parts go, and you should always take care to protect them first. Wearing safety glasses is something to take pride in, a badge of geekdom that you should embrace, heart and soul. Safety glasses will put you squarely on the road to a technical, tinkering future, and you'll most surely want both eyes healthy to enjoy the whole experience.

▶ **Chemicals can damage your skin or eyes, or even catch fire.** There are no desperately dangerous chemicals used in this book, but still: Use caution when you have the slightest doubt about whether a given substance is safe to touch, smell, or mix with other chemicals. Have the fire extinguisher handy and the safety glasses on from the moment you open the bottle of alcohol or bleach, or the moment you begin mixing things you've never mixed before.

NOTE ON THE METRIC SYSTEM

Like most scientists, I'll be using the metric system a lot in this book. In a few places I'll mention miles or gallons or something we U.S. residents are familiar with, but in general, I love the metric system. It's such a brilliant system that it's hard to imagine why we don't use it for everything. The history is pretty interesting. Back in the 1700s, every country and region around the world used different systems of measurement, different units for length, weight, volume, and so forth. Confusion reigned and it was easy to cheat and be cheated.

Then in 1790 five French scientists were commissioned to make a better system, and did they ever. The system they worked out was super simple, super well organized, and rooted in our number system, which happens to be based on the number 10. They said 10 millimeters is a centimeter, and 10 centimeters is a decimeter, and 10 decimeters is a meter, and 1,000 meters is a kilometer, and on and on.

Then, the brilliance continued. These French dudes said, *oui, oui,* the volume should be related to the length, so they put one cubic centimeter to equal one milliliter, and then 10 milliliters is a centiliter, 10 centiliters is a deciliter, 10 deciliters is a liter, and so on. It's the same relations for volume units as for length! Oh, it makes my spine tingle.

But it gets better still. To make the connection between volume and mass, they used one of the most important and ever-present substances on Earth: water. They set one gram as the mass of one milliliter—that's one cubic centimeter—of pure water. And then they linked the other mass units with just the same set of clever prefixes: milli-, centi-, deci-, right on up to kilo-. Isn't it just gorgeous?

You'd think any reasoning person would see this system and snatch it right up, tossing away the old systems like so much rubbish. After all, 2 cups in a pint, 2 pints in a quart, but then 4 quarts in a gallon? What? And 12 inches in a foot, 3 feet in a yard, 1,760 yards in a mile? Gimme a break! You can't be serious!

We got that cumbersome system from the Brits. We sometimes call it the English system, which is kind of a laugh, because the Brits have long since embraced the metric system. (Although they did resist for a long time just because the metric system came from the French, and they hated the French. I like to think the scientists didn't resist.)

We red-blooded Americans, on the other hand, continue to resist. Picture the scene: it's 1977, Ms. Hansen's third-grade class at Eugene Field Elementary School, Maryville, Missouri, time for math. "Children! We're starting the measurement unit today! The United States has decided to convert to the metric system and so we get to learn all these new units of measure. It's going to be great!"

She told us all about meters and grams and liters and explained quite clearly that you don't have to do all the hairy conversion from liters to ounces, or meters to miles, because we're just going to use the new units, and they're a piece of cake. So soon the road signs will all be in kilometers, scales will be in kilograms, and drinks will be sold by the milliliter!

We sat there listening to her with starry eyes; sounded great to us! But back home, the adult murmurs began in many a household: Not going to foist that foreign system on me! What the heck is a kilometer anyway? I've used gallons all my life, is someone trying to rip us off? Is this some kind of a communist conspiracy?

To make a long, melancholy story short, my childhood compatriots and I got the metric system in math class every year for the next three or four grades, measuring stuff with meters, grams, and liters, before it slowly faded away to its current isolated position within the science lab. Out in the streets, we've still got our proud American miles with 5,280 feet, and our proud American pounds with 16 ounces, and our proud American gallons with 16 cups and who knows how many tablespoons, and woe to the person who has to convert between them, or who has to use some metric reference from another country. (Meanwhile, every other nation in greater America, North and South, uses metric.)

The epilogue here is beyond melancholy; it's straight sad. In 1999 NASA launched the Mars Climate Orbiter, meant to gather weather data on the Red Planet. Instead of going into orbit, the $125 million instrument burned to a crisp in the thin Martian atmosphere, because the propulsion system was done partly in metric, and partly in "American," with no conversion. Ouch. I think the engineers cried.

So now you know: embrace the metric system. It's the best thing going.

Part I
Separating Substances

Chapter 1

COCONUT OIL

Get fragrant oil from a ripe coconut.

The coconut is one of the most incredible plants I've come to know. The tree itself is an engineering marvel. It consists of a skinny, serrated shaft that bends and sways with hurricane winds and then stands back up straight, up to four stories tall. The wood has long, stringy fibers running lengthwise that make it useful to fortify houses against a storm. Atop this graceful tendril, a bouquet of lush fronds branch out to form the skeleton of a sphere. These can be used for constructing roofs or walls and the leaves used to weave casings for rice dumplings and other artistic necessities. Nestled under this gorgeous tuft hang the fruits: heavy, round coconuts, full of sweet water and oil-rich meat. If you were designing paradise, you couldn't do much better than this.

The coconut seed and the fruit are intermingled like I've never seen in another plant. Most fruits I know have a big old chunk of good sweet tissue for me to eat, and the complete seed or seeds are inside or scattered throughout. Here the entire entity is the seed. To get a new coconut tree, the densely fibrous husk outside must remain intact until germination occurs. Hack through 4 to 8 centimeters of the massive husk and you get to the hard shell.

If you've never been to the tropics, this shell may be all you've ever seen, because the husks are removed before coconuts are exported. Inside the shell, which is brittle and hard like teak wood, the centimeter-thick layer of rich white meat forms.[1] Inside the hollow sphere created by the meat sloshes the magnificently fragrant and perfectly sweet coconut water.

This is not the milk; coconut milk has to be made from the meat, and we'll do that in the process of extracting the oil. The water is some sort of glorious byproduct from the plant's metabolism, and it will last in good shape for days after the coconut is harvested and stored. Throughout the tropics, young coconuts are sold on street corners for the water to be drunk and the smooth, young, partially formed meat to be scraped off and slurped down. Partaking of one on a sandy beach in Timor can lead you gently, inexorably to the idea that you really don't need anything else in life at all; money, clothes, shelter, and other food and drink fade away to distant priorities in comparison to the blissful crystal of celestial perfection that you are currently imbibing.

I digress. Coconuts are one of the few local sources of oil for the Timorese, so they are highly sought after and the trees are well tended. At some point, it hit me that each of the thousands of coconut palms you see traveling around Timor was planted by someone and is owned by someone. Of course, they can reproduce naturally, but if that happens here, someone jumps on the sprout and claims it for his or her own.

MAKING COCONUT OIL IN BUKOLI, TIMOR-LESTE

I watched this process 15 years ago in the mountains near Baucau, Timor-Leste's second city. Our friends there have done this for generations and were happy to let me photograph them with an early digital camera. I found it took all afternoon and involved breathing of a lot of smoke. I missed seeing the part where they grated the coconuts. The girls with the gorgeous smiles in the photos are now finishing up university.

Here they're loading the grated and soaked coconut into the lever press.

The family takes turns sitting on the lever press, waiting for the milk to trickle out.

Now they're cooking the coconut milk.

Here they're carefully separating the oil from the candy once it's done cooking.

Now they're loading the candy into a bag, tying it off, and squeezing it with the giant nutcracker.

Finally, they get to enjoy the candy.

To get the oil, you have to let the coconut grow to maturity, dry, and turn brown. You can either chop it off or let it fall on its own (watch your head—and your car!). Then the arduous process of extracting the oil from the meat begins. Let's do it!

Gather stuff

▶ One or more whole ripe coconuts

▶ Water

▶ Sweet potato, if you have one

Gather tools

▶ Hammer

▶ Knife

▶ Chisel

▶ Grater

▶ Bowls

▶ Cloth strainer or clean rag

▶ Stove

▶ Wok or frying pan, nice and thick if possible

▶ Spoon

▶ Ladle

▶ Tongs

▶ Jars or glasses

TINKER

Your first task is to open this bugger. You can try slamming it onto concrete or hitting it with a hammer. You could saw it in half, but then it's hard to get the meat out. If you can feel any water inside it, try to save and taste it; it's often not bad, but not as good as young coconut water.

Now use the hammer and chisel to reduce it to smaller shards so it will be easier to get the meat out. But you don't want the shards too small, because you'll have to hold on to them as you grate them.

Now you have to separate the meat from the hard shell. This is a real trick. We used a knife and chisel. The Timorese often use a machete to hack away at the shell from the outside, but I figured you might not have a sharp machete—or know how to use it for this.[2]

Here the chief challenge is to avoid stabbing your hand. Be sure to stab and chisel away from any of your body parts. You want to work hard to separate the meat from the shell in large pieces,

because if they're too small, you won't be able to grate them.

Once all the shell is separated from the meat, start grating. Use the finest section of the grater, the one where the bits don't go through to the other side but fall off down the front. You want to grate it into the smallest pieces you can, to get at as much of the oil as you can. If you start grating your fingertips off, imagine if this were your sole means to attain cooking oil.

You don't want to waste all those little bits you couldn't hold on to anymore, so put them on the cutting board and mince them the best you can. Every cut is important to liberate more of that stubborn oil from inside the dreamy white meat.

Halfway through the last time I did this I realized that our blender was sitting in the cabinet to my left. After a millisecond of hesitation, I hauled it out and did the rest with electric power. I can tell you that it still took a long time, because the blinking blender didn't work with dry coconut, so I had to add the water—the next step in the process—which made it a mucky mess. You don't really want it blended; what you need is a superfine chopper. But I didn't have such a fancy thing. The blender did end up beating a heck of a lot more oil out of my coconut meat than when we do it by hand. Of course, all this electrical stuff is cheating, but I guess it's up to you in the end.

When it's all grated, add a bit of hot water, then mush it all together. Every step in this process is designed to encourage the oil to emerge out of the pulp. The commercial professionals add just a tiny bit of water until it's up to some exact wetness proven to be most effective at releasing the oil, and then squeeze it with

massive hydraulic presses. It's not easy to get such technology, so just know that every bit of water you add will have to be removed in the end. We added slightly over one cup of water for our average-sized coconut. Then we mushed it together for a while.

Now you need your cloth strainer. These are ever present here, because there are many things the Timorese use just the juice from. If you don't have one, you can get a tight piece of cloth, maybe a handkerchief, or you might cut out the pocket of some discarded pants. Socks are the right shape, but their weave is a bit too loose (plus, think where they've been—eeww).

Get a bowl ready for the milk and then put a good heap of the coconut mush into the strainer held above the bowl.

Now squeeze the bejeepers out of it. (Check out how they do this in the mountains, in the photo essay in the previous sidebar.) Every drop of liquid coming out of this will add a bit more oil to your final product, so do your best! We use the twist-and-squash technique.

Dump out the dry remains when you've squeezed your best.

After a few rounds of straining and squeezing, there is your luscious coconut milk. And there are the forlorn gratings, dry and oil-less.

Or are they? There is an ongoing debate among my teacher colleagues about whether it's worth it to splash a bit more water onto these, mix them up, and squeeze them again. It may release a few more drops of oil, but boiling the water off will then require more time and fuel. Let's forget about it for now.

COCONUT MILK

At some point in this process, you'll undoubtedly get to thinking about whether it's really necessary to go through this whole rigmarole. Because, hey, the oil is inside the milk, even though it's not clear like other oil you've used, so maybe you could just take this coconut milk here and cook with that!

Good news: you can. Coconut milk is a common ingredient in household cooking here. You should be able to find it canned or boxed in the Asian section of your local supermarket. If you've ever had any sort of Southeast Asian meal with sweet, yellowish rice or sauce, this is what it was made from. It's a no-brainer from both culinary and practical standpoints. The only drawback is that it doesn't last long. Without refrigeration or proper preservation, this stuff will go bad in a couple of days. That's why ripe, raw coconuts are always hanging around houses here. Just like the ones you may find in your supermarket in the temperate climes, ripe coconuts will keep for months in that semi-dry state, until you decide to make either oil or milk from them.

Here is where the paths part for virgin and non-virgin oil. You can try both, even with only one coconut. Divide your coconut milk into two parts: one part into a wok and one into a glass jar.

Place the glass jar on a shelf for a couple of days. That's all there is to making virgin coconut oil: patience and then a bit of finesse in separating it out. It will start to separate immediately. The following photo shows the jar after 15 minutes.

For the process where you boil the milk—the process nearly everyone here uses—put the wok on a burner and set it to boiling on low heat.

Now get comfortable, because this takes a while. Basically, you've got to wait until all that water you put into the gratings boils away. The smooth milk will soon start to take on a curd texture. You can stir it if you don't have anything better to do, but you don't really need to. Just keep an eye out for the first little pools of oil to appear among the curds of coconut milk.

Here things begin smelling heavenly, something between a Malibu beach and a coconut macaroon. You begin to realize that there is some serious sugar in there as well as the oil, and when it caramelizes with the heat, a lovely scent is sent up.

When you finally see little pools of oil, you can try to grab them with the ladle or a spoon. Plop it in gently and try to get the oil to flow in over the side without getting any of the curd crud. Some people keep this early oil separate. It's cleaner and clearer, and it will keep for longer than the cooked oil.

But you don't have to do that, especially with the piddling amount of oil you get from an amateur job on a single coconut. You can go straight to the next step.

As soon as the majority of the curds look hard and crispy, take the wok off the fire. If you leave it on a couple of minutes too long, the oil and curds will burn and you'll have ruined them, so don't do that. Now all that's left is to separate the oil from the crackly remains of the curds.

Wait a few minutes, because the next step is to work with the stuff that was just boiling. There is not much heat to dissipate, but be sure to wait at least 5 minutes until it cools down a bit.

Bring out your cloth strainer again. Wash it off well, and dry it too, the best you can. It's not good to have water in your cooking oil. When you think the oil is at a workable temperature, put the strainer into a bowl and dump the contents of the wok into the strainer.

Pick up the strainer and squeeze it. For more force and to avoid getting too oily, we used tongs like a nutcracker, but maybe a nutcracker would work even better. You have to keep the strainer tightly closed and then squeeze it every which way as you watch the golden oil drip out into the bowl.

Again, no self-respecting coconut farmer will leave this part of the process until every last drop of oil is squozen from the candy in the strainer. Ah, yes, and it's candy that's left in the strainer when the squeezing is done. Have a bite.

Rich? Like, the richest thing you've ever sunk your teeth into your whole life long? I find it so rich, sweet, and greasy that I can't take more than one spoonful. I feel like I've eaten a super-concentrated donut.

Now pour your oil into a jar and smile. Here is the result of all that work. We got about half a cup. Wow.

The virgin oil will take a day or so to separate out.

This gives you time to figure out how to get the oil away from the other stuff in there. We usually get a nice, milky water layer on the bottom, and the oil is still mixed with some curd-like stuff up top. The bad news is that if you dump the oil and the curds into a strainer, they both go through! Here is what we do: pour the oil and curd off the top, careful not to get any of the water from below, though inevitably some goes in. You can help it along with a spoon.

Now use the ladle or spoon method to selectively dip off the pure transparent virgin oil. Try to avoid all the water at the bottom.

When you're down to the last bit, you can either put it in the pot and cook with it on the spot or try to pour it through the strainer again, or an even tighter cloth, like a cheesecloth, to try to separate the curds. Don't squeeze it this time, because the curds come right through the cloth.[3]

And there you have it: virgin coconut oil.

Together with what you cooked up, that's the sum total of all your efforts. Whatdya think? Worth it or not?

Wait, wait, first grab a sweet potato, slice it thin, and fry it up with one of these oils. Then reconsider: Worth it or not?

MACHINES DOING THE DIRTY WORK

Nowadays in Timor, you can get the most strenuous step done by a machine at the market for a few cents a coconut. I checked out the one at the market near my house.

The coconuts in the pile to the left have been shorn of their hard shell. This is an enormous trick requiring a sharp machete. You can see where the blade cut through the brown membrane and into the white meat. The man is chopping my two selected coconuts into smaller pieces.

He then feeds them one by one into the gaping mouth of the shredder. Powered by a gasoline engine, it rotates a grater drum at a scary speed as the man nonchalantly pushes the pieces in with his fingers.

At the very last instant he pulls back his fingers and leaves the last bit to rattle around for a while. You can see this is one machine that never needs oiling, since it's daily bathed in oily coconut milk.

When the two coconuts are grated, the man gathers the results in a plastic bag.

And I'm ready to go. Total time: 2 minutes.

On the way out, I notice bottles of coconut oil for sale with corn and beans in another stand. They were imported to the capital in spent drink bottles. It's all been cooked; no virgin available.

WHAT'S GOING ON?

As with other oil-producing plants (sunflowers, soybeans, peanuts, etc.), the metabolism of the coconut results in an oily tissue, coconut meat. The plant uses this tissue to store away energy for use when it germinates and begins to grow into a new plant. And as in the case of all edible plants, we step in and take advantage of this clever little talent. None of the other plant sources of oil can give as much from a single fruit.

Who would have thought that those wondrous orbs hanging up under the coconut tree tuft would be able to produce oil? It must have been quite exciting to the original societies in prehistory who figured this out. In fact, research suggests that nearly all modern food crops have been modified by systematic management over generations of early farmers to result in higher yields of better nutrition than the original plant provided.

Here in the tropics, various critical crops come from trees. Aside from coconuts and palm kernel trees providing essential oil to the diet, jackfruit, breadfruit, bananas, and papaya are all used as either vegetables or starches, some becoming the main caloric intake during parts of the year when other local foods are not available. Then there is the sago palm, the very trunk of which can be eaten if you process it right. In temperate zones it seems to me that trees are less critical food-wise, sort of the dessert of the diet: fruits and nuts and maple syrup. It's been great to learn so much about the way people make use of trees here in the tropics.

COCONUT HARVESTING

Getting at these nuggets of goodness is a bit of a trick. You can just wait for them to come down, but that may take a while, so it's better to go up after them if you want the young ones. Here are some photos I took near a house we were staying at when a guy showed up to harvest the young ones to sell for drinking on the streets of Dili.

Don't be fooled by the rope—it's for the coconuts, not his safety.

Steps have been chopped into the side of the trunk to ease the climb.

The terrifying part is grabbing the branches and pulling up onto the top of the tuft, now easily 20 meters—60 feet—off the ground.

After working out a path for the rope, he's drawn his cleaver, ready to chop loose the clump he'll lower first.

The guy standing under the bugenvila bush below slowly lets the coconuts down on the rope.

They've got the first bunch down safely.

After lowering several more bunches, our hero descends.

"Look mom, no hands! And a cleaver to boot!"

THE PERILS OF PALM OIL

Most of the cooking oil used in Southeast Asia, and a large fraction of the oil used in processed, packaged foods worldwide, is not coconut oil but palm kernel oil. The oil from nuts of the *Elaeis guineensis* tree give an oil that is less fragrant, more neutral, and stable for longer than coconut oil. Palm kernel oil is good business, and as the world's population increases, companies looking to plant more of these palm trees are chopping down vast swaths of rainforest, some of which are one of a kind, and some of which are the sole habitat for orangutans and various other species.

CREDIT: BRETT JORDAN

CREDIT: DAVID STANG

Palm kernel oil is such big business that it has often overruled land rights and environmental protection laws in tropical countries. Various international organizations are working to make palm oil more sustainable and protect endangered species, but as I write this, there is no sure way of getting palm kernel oil that was farmed in a sustainable, environmentally friendly way. In fact, some of the organizations putting a happy eco-friendly stamp on bottles of oil are in fact funded by the companies themselves to give a deception of goodness. It's a conundrum that we in the United States can somewhat avoid by buying cooking oil produced locally.

ENDNOTES

1 Friends here laugh when we translate this directly to Tetun; "meat" is definitely not something that comes from a coconut. But their word for the chewy white stuff—*isin*—is equally funny to me. Directly translated, it means "the body," whereas actually it's just a layer of stiff tissue. A more meaningful translation would be "essence," because this body is the key element, where the luscious oil is hidden away.

2 Formed from old truck springs on forges blown by bamboo bellows, Timor's machetes are truly a tool for all jobs. Aside from preparing coconut oil, the Timorese can build entire houses with no other tools beyond the machete.

3 I saw on the web how you can put the mixture in the freezer to help along this part of the separation, because the water will freeze and the oil and curd won't. It's a bit of a trick to do that if you don't have a freezer, but I assume you do, so try that if you're not a purist.

Chapter 2

REMIXING AIR

Light a candle in a bottle to change the composition of air.

Sometimes linguistics can shed light on science. In Tetun, there is no equivalent for the English word "air." There is a word meaning wind—*anin*—but it's clear that this word does not describe the contents of an empty bottle; *anin* has to be moving. The words meaning smoke (*ahi-suar*; that's wood-*suar*) and steam (*bee-suar*; that's water-*suar*) don't mean air, because everyone agrees that if you can't see it, it's not *suar*.

Tetun has taken a lot of words from Portuguese over the last 500 years, and so most educated people know both *vapór* and *ár*, the Portuguese words for vapor and air. Still, I'm always curious to think about what words the precolonial Timorese of a thousand years ago used to describe, say, what's in your mouth when you close your lips and puff out your cheeks, or what's inside bubbles in a frothy soup, or what stands in the way of a leaf fluttering down from a tall tree.

Trying to describe or analyze this ever-present yet invisible substance sets you looking for the tools of science. You can read details about the fascinatingly simple experiments the British scientists Mayow and Black and Priestley used to understand air in the 1600s and 1700s, but you can also be sure scientists around the world and throughout history have stretched their investigation strategies as well as their vocabularies to grasp the gist of air.

In round numbers, the textbooks say air is 78 percent nitrogen, 21 percent oxygen, and 1 percent various other gases. Immediately there is a caveat the textbook may or may not mention: water vapor is also always part of air, and varies from 1 percent to 3 percent. This unavoidable, damp detail skews those other numbers *in all cases*. So much for that pert little textbook fact so many millions of students are made to memorize. As usual, it's "right," but it's more complicated than that; you can't *not* mention that those percentages are for the non-water components of air.

We'll get a combustion reaction going in a closed space and see what we can deduce about the air that fills that space.

MY CLOSE CALL WITH COMBUSTION

Combustion reactions—that is, fires—have been the cause of much death and destruction over the centuries. If you include fire in your tinkering, you'd best have a plan for putting it out should it spread. As a kid, I once had a tissue paper hot-air balloon suspended in my room, powered by a ketchup bottle lid full of gasoline—terrible idea, by the way. It was my great fortune this was not one of my secret projects; my dear old dad was brave enough to let me try it, and wise enough to grab the fire extinguisher.

Upon ignition, great hot flames licked up into the bag full of air and began to set the whole thing rising, but also swaying, which splashed a bit of the gasoline out onto the supporting balsa wood strip. The balsa burned through in no time, dashing the flaming liquid all over the linoleum and nearby rug, creating a toasty wall of fire in the midst of my bedroom. My dad calmly pulled the pin on the extinguisher and blasted its fine yellow powder across the scene. As the smoke and dust slowly settled, I remember him drilling me with his eyes and saying, "Never again."

WARNING Moral to the story: *don't* play with flames, not even candles, without a wise adult around, and *do* think about what will happen should you lose control of your combustion. Oh, and the bedroom is a lousy place to do combustion experiments

Gather stuff

- Bottles and jars, glass and plastic. Various sizes and shapes will add more insight. Plastic ones may melt or catch fire, but glass ones may drop and break, so either way, beware.
- Candle that fits in the neck of the bottle
- Matches or a lighter
- Fire extinguisher
- Plate or bowl
- Water, with coloring if possible
- Optional:
    Large magnifying glass
    Steel wool
    Thin wire

Clip leads or thin wires

Batteries or a battery charger

Cardboard

Rubber bands

Mirror

Hot water

Cold water

Aluminum can

Gather tools

- Scissors to cut hole in bottle
- Optional:
    Wire strippers or a knife
    Burner
    Tongs
    Basin with water

TINKER

The basic idea here is to light a candle and then put a bottle over it. You can just stick a candle down with a bit of its own wax, or you can make a snazzy wire stand. We do this on a glass plate to reduce the chance of unplanned fire.

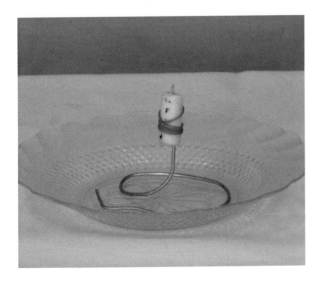

Light the candle and then lower the bottle over it.

By golly, it went out. Raise the bottle a bit, light the candle, and try it again.

And again. And again. Keep going if you think you can observe something with more tries. That goes for all the following activities.

Variation #1: Put a Hole in the Top

For this one, you need a plastic bottle. Put a modest-sized hole in it and try it again. Put a smaller hole or a larger hole and try those too. Don't let it go too long or the plastic will start melting.

Variation #2: Put a Lid on It

Now instead of sitting the candle on the plate, sit it in the upturned lid.

Put the bottle on and twist the lid shut tightly.

What we see is a self-squashing of the bottle as the candle goes out. Hmm.

Variation #3: Put it in the drink

Now for the main event! Stand the candle up again and pour water into the dish.

It's easier to see what happens if you put some color in the water.

Light the candle and cover it with the jar. Watch very carefully everything that happens.

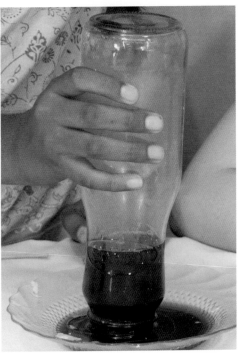

The mysteriousness of water being drawn into an upturned jar when the candle is already out is enough to make my teeth itch.

WHAT'S GOING ON?

This is a common experiment and many sources wrap it up with the following conclusion: "And thus you have proven that air is around 20 percent oxygen." Balderdash. What you've "proven" here is small and subtle.

It never fails to draw a laugh and a puzzled sigh when a student lays out the balderdash conclusion and I hasten to their experimental setup with a shocked look, saying, "That's incredible! Where is the oxygen? I've never been able to see it before!"

You can't see it, and darn good thing: it's all around us, so if you could see it, you couldn't see much else. You can't actually see many gases. My local university chemistry department set up a display showing mercury boiling at 357°C for my eighth-grade class—a metal changing from liquid to a gaseous state, which is something you don't see every day—but we were disappointed because all you could see was the shiny burbling liquid on the side of the strange glassware over the high-powered stove, and then a calm pool of the same silvery stuff slowly condensing on the other side, where it was cooler. This is the consternating characteristic of invisibility that has been such a challenge to scientists in the course of figuring out air.

In fact, it's been challenging all the way to the present day, when you can still find plenty of sources explaining this experiment incorrectly or incompletely. It's very tempting to start with and focus on explaining the water variation (#3) because it's so cool and unexpected to see the water being pulled up into the bottle. But to understand what's happening, you'd better do and explain the first two first. And then you'd better set up the additional ones I'll explain in a moment.

With the original activity, covering a candle with a jar, the key to understanding is actually in the attempted *repeat* of the activity: lifting the bottle, relighting the candle, and lowering the bottle over the candle again. If yours works like ours, the candle goes out the instant it reenters the bottle. Lift, light, and lower again: again, there it's instantly out. Again and again you can do it, and you'll never get the same long burn as the first time.

Unless you change the air in the bottle. You can do this by waving it, blowing into it, squashing it a few times if it's plastic, or blowing into a straw stuck into it. Then when you light and lower it again, you get several pleasant seconds of candle light until it goes out again. Be sure you try this yourself.

So the conclusive result of the basic version is that when you let combustion happen in a closed environment, it won't last long. Once it stops, the air in there is not the same—specifically, the fire can no longer get what it needs to burn. So the flame is doing something to the air in there, such that after a few seconds, the air in the bottle is no good to support more flames.

That's all you can conclude! Say any more and your pants are on fire!

Scientists over the last 300 years have worked out that the part of air needed to burn something is called oxygen, and that's what the candle is consuming until there is not enough of it to support burning. But those masters had better labs and more time. All you've got is this junk on the table. So feel free to expound on the scientist's findings about the composition of air, but make a clear distinction: we can get a feel for the truth of those findings from this experiment, but what we can really conclude with our current setup here is severely limited.

But really, this is pretty cool. The air may look the same in there, but it's now different. Put more air in, and the candle will burn again. Put a hole in the top of the bottle (variation #1), and the air can keep on circulating through, so the candle gets what it needs to avoid dying.

To explain the other variations, we have to bring in a few other astonishing facts, which can be shown with various other simple demos. I hope I needn't remind you to actually try these yourself and don't just take my word for the results.

First, when gas gets hot, it expands. Put a balloon over a bottle mouth and then pour hot water on the bottle. The balloon perks up because the air in the balloon-bottle system is expanding with the heat.

Second, when gas gets cool, it contracts. Take the balloon off the mouth of the bottle that's sitting in hot water, squeeze the air out, attach it firmly again, and then bathe the bottle in cold water and ice.

As the air in the bottle contracted, the balloon got sucked into the bottle!

Third, one of the products of combustion is water vapor. Find a small piece of glass, like a pocket mirror, and hold it briefly over the flame of a lighter or candle. (Not too long or you may crack the glass! And make sure you're not dealing with a plastic mirror!) You can usually see the glass get a small haze of condensation just outside the area hit by the flame.

It dries off quickly because it also gets hot, which makes it tend to evaporate again. But this haze is the water being produced by the lighter's butane or the candle's paraffin combusting with oxygen in the air, and the important thing is that it is produced as gas and then condenses to liquid on the cooler glass.

Fourth, and most critical to understanding this activity, when water vapor gets cold it condenses into little droplets. When that happens, it decreases in volume 1,000 times. This is the source of some of high school chemistry's finest demos. Witness the imploding can.

Heat 50 drops of water in an aluminum can over a burner until it's boiling well.

Invert the can immediately into cold water. Zump! One squashed can.

Screw on the top and watch the fun begin as it cools.

The slow-motion version heats the same 50 drops in a can with a screw top until boiling strong.

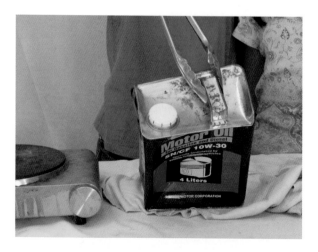

Paul Doherty used this as metaphor for students who didn't do their homework, even though it didn't count for much credit. You'll be steaming along fine thinking you understand the material, but you'll begin to feel like this can at exam time.

By the way, it turns out this 1,000× volume quick change with the shift in states of matter is good for more than just wowie-zowie science demos. It also gave force to the Industrial Revolution about 300 years ago. A steam engine takes advantage of the increase in the volume when water turns to gas.

Now put these four facts together to explain variation #2, the capped bottle:

▸ Candle produces heat and water vapor with its combustion.

▸ Heat expands the air in the bottle and, together with the produced water vapor, pushes some air out the mouth *even before* you get the lid screwed on.

▸ Water condenses on the sides of the bottle (can you see it?) after the candle goes out and the bottle cools, reducing that part of the volume of the bottle's air 1,000 times.

- Air in the bottle also cools, reducing its volume by a smaller ratio.

- Atmospheric pressure, still pressing away as normal on the outside of the bottle, crushes it a bit to make up for that lost volume on the inside.

  Now variation #3, the rising water show:

- Candle produces heat and water vapor with its combustion.

- Heat expands the air in the jar and together with the produced water vapor pushes some air out the mouth, making bubbles in the bowl of water. (Did you see them?)

- Water condenses on the sides of the jar after the candle goes out and the bottle cools, reducing its volume 1,000 times.

- Air in the jar also cools, reducing its volume by a smaller ratio.

- Atmospheric pressure is still pressing away as normal outside the jar and the surface of the water, so the water is pushed up into the jar to make up for that lost volume.

COMBUSTION EQUATIONS

Let's look at the chemical equation that represents the combustion reaction. In essence, combustion divides the air and reconfigures its molecules using some carbon and hydrogen atoms from the fuel. Specifically, this candle flame pulls the oxygen out of air and replaces it with carbon dioxide and water vapor, the latter of which will then condense when it cools. Don't forget that you're going to need a much bigger lab before you can actually prove that.

Here's an ultra-simplified chemical equation, using a formula for a hydrocarbon similar to paraffin. It's a reasonable representation of what's happening in there:

$$\text{Oxygen} + \text{Combustible} = \text{Carbon Dioxide} + \text{Water}$$

That's

$$O_2 + C_2H_6 = CO_2 + H_2O$$

Or, balanced:

$$7O_2 + 2C_2H_6 = 4CO_2 + 6H_2O$$

Now look at the physical states of the compounds in our experiment:

$$\text{Gas} + \text{Solid} = \text{Gas} + \text{Liquid}$$

Actually the paraffin melts to a liquid and boils to a gas in the candle wick before it hits the combustion reaction, but its origin in the experiment was the solid candle. And the water from the combustion reaction was produced as a gas, but then condensed when it hit the cold wall, as we've noted. Thus, long before the reaction started, the oxygen gas and the paraffin solid were waiting at room temperature to react. Also, long after it ends, the carbon dioxide gas remains and the water is in little liquid droplets at room temperature on the side of the bottle.

So, a gas and a solid are converted to a gas and a liquid. The liquid and solid states are nearly the same volume for most materials; it's that gas state that goes through the roof at 1,000 times the volume. With this basic info, you could predict that air will blast out of the jar's mouth when the candle enters it, and then water from the dish will surge back in as the water condenses and the air cools.

Let's take a look at the explanation fallacies that pop up so often. A few sources I've seen simply say (falsely) that the oxygen is "used up," dematerializing so to speak, and the water flows in to take its place. The authors (and editors) of these sources need to spend some quality time with Uncle Lavoisier, who said (after completing a great many experiments in the 1700s) that matter is never gained or lost but rather changes form. In other words, that oxygen ain't gone, my friend; it's right there in the CO_2 and the H_2O.

Some sources confidently state that the reason the water rises is because carbon dioxide is produced in combustion (true) and carbon dioxide is more soluble in water than oxygen (also true: 20 to 30 times more soluble). Thus, when extra carbon dioxide is produced, it gets quickly dissolved into the water, which rises to make up for that lost volume.

Well, our variation #2 doesn't involve any water at the base at all, but the bottle still gets squashed. Hmm. Another way to disprove this one is by replacing the water with oil, which should have a different solubility but gives a similar result. Try it!

Finally, a few sources, including big guns like Bill Nye and Steve Spangler, don't mention the

state change at all but blame the water rise on the air heating and cooling. I'm convinced that simple expansion and contraction with temperature can't explain all the bubbles or the rise. I base that on the balloons' movement in the earlier demos on the ketchup bottle with hot and cold water. Also, the timing of the surge of water doesn't seem to follow the temperature change, though it does seem to mimic the imploding can and other state change demos I've seen.

But the best way to understand this extraordinary little experiment is to figure out a way to light the candle *already inside the bottle!*

Variation #4: Light the Candle from Outside the Bottle

It helps to arrange a tropical sun to make this happen. If you've got a good, powerful sun, you can light the candle at midday with a magnifying glass through the side of the bottle. Wow.

By the way, Mayow used this same technique to heat an enclosed substance in 1664, but I still say I discovered it, because when I thought of it here, while sweating away an afternoon in the lab, I hadn't read about him yet! It's amazing how often the same thing can be discovered!

We found it hard to light the candle wick directly, so we learned to mount a match just beside the wick, and then use the magnifying glass to ignite the match head, which then ignites the candle wick.

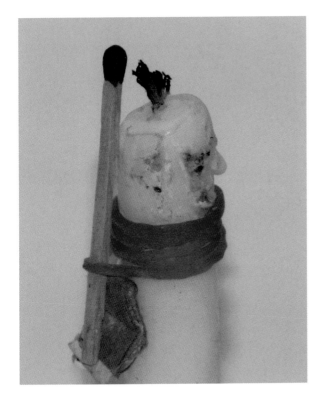

If yours works like ours did, you're sure to see plenty of bubbles upon ignition, and then the water rises to the about same level or higher than it did in variation #3.

By the way, if you don't have access to the tropical sun, you can use a model rocket ignitor for your candle, or use steel wool to make your own electric ignitor around a match, like this:

We put the rubber bands on first, and then jammed the match in parallel to the candle. Note the little nub of cardboard at the bottom that gets the match angling over near the wick.

If you hold a piece of paper at the side of the bottle, you can gauge the approximate distance to get the focus right at the match head.

It's not super easy, because the glass of the bottle also bends the light, making the focus in a slightly different place, and it also absorbs some of the radiation that you are wanting to concentrate in order to light the candle. But with enough sun—and mercy, do we have plenty of sun here—it lights.

I'll let you work out the details, but here is a photo of ours, just before ignition.

Variation #5: Suck In Some Water First

This one is the real kicker: air never escapes from the bottle, but you get to watch as the water level changes back and forth. Stick a small tube or long straw up into the bottle as you lower it over the candle.

Now suck out some air from the bottle and watch water move up into the bottle to take its space.

Quickly withdraw the tube, leaving the water level high up the neck.

Then light the candle again and see what happens.

I couldn't capture the process with a photo, but "inarguable" is the word that comes to my mind. We saw the water go down, then come slowly back up to the same position. As with most paradoxes, once you've got the right perspective and enough information, there's no paradox at all.

If you're still not tired of this one (as I'll never be, long as I may live), you can try two more variations:

Variation #6: Different Bottles Get You Different Results

Remember your science: to see the influence of one variable, you have to keep all the other variables constant. So make sure you have three candles exactly the same size and try them beneath three different-sized bottles.

Variation #7: Different Candles Get You Different Results

Now change the variable: get three bottles exactly the same and arrange three candles of different heights.

What did you discover? What other experiments can you think of to shed more light on air (without burning the house down)?

Chapter 3

BIOGAS

Make and burn methane from garbage and your dog's droppings.

What I if I told you that you could cook something using your kitchen garbage and your dog's droppings? It's true! People all over Asia and Africa are doing it with tanks out behind their houses. You can do it too! Actually, with the rig I'm going to show you, you may not get enough energy to do more than toast a marshmallow, but you will see how it's done and get a nice warm blue flame, and that's always interesting. Before we begin, let me point this out that getting energy from garbage makes perfect sense from the perspective of chemistry. Most things that burn are hydrocarbons, meaning they're composed of molecules rich in—you guessed it—hydrogen and carbon. Well, practically everything we eat is also hydrocarbons. Life is hydrocarbons and we depend on life—mostly plant life—to live!

When bananas go bad and they're no good to eat anymore, their hydrocarbons are still in great shape. Also, unavoidably, when we make that trip to the toilet after a satisfying day of digesting several meals, what comes out is—that's right: hydrocarbons. This is true for all animals, and many other forms of life as well (except for the part about the toilet).

So it makes total sense that you could somehow collect those hydrocarbons that you don't have any other plans for, process them in some clever way, and burn them for cooking fuel. In many parts of the world, people gather cow dung and dry it for cooking fuel. Cows eat grass (hydrocarbons) and from it produce massive quantities of stinky manure (hydrocarbons).[1] But burning cow dung probably won't work so well in your kitchen,

35

what with the smoke and the smell and all. What we need is a nice clean gas.

Well, that's exactly what we'll get here: methane. The experiment will only stink when you open it to add garbage. Aside from that, you could keep it in your living room, beside the potted plant.

Gather stuff

▸ Two containers that fit together, one upside down on top of the other. The bigger these are, the more biogas you'll produce, but the more garbage you'll have to feed it.

▸ One more vessel, like a basin, which can be filled with water and which the first two containers fit into

▸ Small flexible tube, 2 meters

▸ Something to plug the tube with. We use a round pencil.

▸ Aluminum foil

▸ Safety glasses

▸ Matches or lighter

▸ Water

▸ Duct tape

▸ Garbage and animal droppings (or just the droppings, if you can get enough of them)

▸ Marshmallow on a stick, or something else to cook. (Um, I hope it goes without saying that you don't want to get this item too close to the biogas rig until you have a flame burning...)

Gather tools

▸ Scissors

▸ Shovel for collecting the droppings

▸ Stick for stirring the garbage

TINKER

First we'll make the rig. The garbage and dung will go into the small container, and the bigger one will go over the top of it and be sealed with water around the bottom so the gas doesn't get away. The tube will be in there to allow the gas to escape. This rig is called a "digester."

Tape the tube to the side of the small container, with its tip just at the top of rim.

Bring on the garbage! Fill the container about half full with nice, wet, rotting garbage: banana peels, bread, rice, oatmeal, pasta, veggies, and fruits. Eggs, milk, and meat are okay in small quantities.

Now for the dung: droppings from dogs, cats, birds, rabbits, guinea pigs, and certainly any farm animals you may find standing around. Here in Timor we usually use straight cow or pig manure, no garbage, and it works great. But I was thinking maybe you don't know too many cows or pigs personally, so I tried it with just dog doo-doo, and it still worked.

Most of my neighbors have dogs, and they relieve themselves in the street each night, so I just had to get up a couple of mornings in a row, before the neighbors went out to sweep their section of the street, and nab the good stuff. Our biogas digester worked fine with half the container filled with garbage and topped with around 10 steaming dog droppings.

WARNING Don't use human waste! Many serious gut sicknesses arise when the bacteria present in our solid waste (poop) get back in through our mouths. Even with the animal waste, wash your hands well after adding it to your bucket. Professional operations can and do use human waste, but we're not going there for safety reasons.

Now fill the rest of the small container with water and stir it up, baby! The better it's smashed and mixed up, the better it will work. I saw a website showing a guy that puts his through an old garbage disposal. What you're looking for is small pieces easily available to any passing microbes.

You're also looking for an anaerobic environment—that is, a situation where the microbes in all that oozing garbage and dung don't get oxygen, or at least not much of it. Thus, you soak it in water. Now only the top surface can get oxygen, and since we'll cover it with the other container, that oxygen will soon be consumed by other microbes and the whole thing will be anaerobic.

Tape the other end of the tube to the edge of the big basin. This will be the place you light the flame. You can also see we put a collar of aluminum foil around the tip of the hose. That's because once it's burning, the plastic of the tube gets too hot and starts to melt. The aluminum collar extends a bit up from the tube so that the flame is away from the plastic. This time we used a tight-fitting pencil to plug the hose.

Now put the garbage container in the basin with the tube going down and coming back up and out the side. (It looks like the water spinach stalks all floated to the top; we should have chopped them up more.)

Fill the basin with water. This water will be cleanish, so try to keep the slop from the other container from splashing out into it.

And finally, with the plug out, put the other container over the first one, pushing it slowly all the way down as the air escapes out the hose.

When the big container is down all the way, plug the end of the hose tightly. This is so the methane doesn't leak out until you need it. It needs to be airtight, so jam the plug in well.

Are you ready for the next step? It's a killer: put it in a nice warm spot and wait a few days.

But you don't have to sit there keeping it company. Those little anaerobic microbes are happy enough working on their own. You'll want to check it every day, but how do you know if anything is happening?

When the ooze is producing methane, it will be bubbling off the top of the slop bucket. You may not be able to see these bubbles, but they'll collect under the larger container, which is positioned to float up as the methane is collected. So if your rig is working, the upper container will begin to rise after a few days.

When the top container has risen a bit, you may want to try to burn the methane, but it's best to just get rid of this first gas. It's mixed with the air that was in there when you put it on, which means it will still have some oxygen, and so it may burn inside instead of out the end of the tube like we want. Plus, in my experience, some other gas or gases are generated first, and they don't burn like the methane we're looking for. So just get rid of them; unplug the tube, push the container back down to expel all the gas out from inside, and then put the plug in tightly again.

Sometimes biogas operations explode, but usually they don't because pure methane alone can't burn. It needs oxygen, like most combustion reactions, and if all goes well, there is no oxygen inside the tank. As it heads out the tip of the tube, it encounters the oxygen in the air around us and forms a nice, burnable mixture. This is what happens at every gas-burning stove, butane lighter, Bunsen burner, or propane torch. It's not what happens with an acetylene welder; there you've got a separate tank full of oxygen, which gets blasted in together with the acetylene and makes a super-hot flame.

When the top container rises again, as it surely will in another day or so, you can try to burn some methane. Put on your safety glasses, light a match, hold it near the tube's end, and remove the plug.

You can't see the flame here because of the sunlight. It is a not-so-bright blue flame hot enough to singe the hair on my hand and toast a marshmallow, but sometimes I can't see it at all. At night it is beautiful. By contrast, the flames from pure pig manure are deep yellow, almost sooty. Not sure what's going on there.

If you can't get a flame going, push the top container down a bit, make sure gas is coming out the tube, and try to light it again.

If nothing lights, no worries—sometimes ours have taken more than a week before getting those little microbes cranked into high gear. Sometimes the top container rises right away and then rises every day afterward, and other times it takes several days to rise after a push down. Some of it depends on the temperature. If it's under 60°F where you are, you may want to arrange some way to keep your bacteria warm in there. I notice when it's nice and hot here, like 95°, it tends to produce gas faster.

After a couple of weeks, if nothing has happened, or if the top container rises but the gas doesn't burn, perhaps you'd better open it up, take out some of the garbage, and put in some more droppings.

When it does start burning, stick the marshmallow in there! We couldn't find any marshmallows in Dili (I know, it's hard to understand how a city of 200,000 can function without access to marshmallows...), so instead we roasted a tiny fish kabob, called "satay" in Indonesia as well as here in Timor.

TROUBLESHOOTING

If your container doesn't rise:

▸ Check for leaks. Is your plug secure? Is there a hole in the container? A hole in the hose?

▸ Add heat somehow. Put it in the sun or inside a warm porch.

▸ Open it up and stir the garbage again. Use a long knife or something thin to make small pieces, all covered with water.

▸ If you do all this a couple of times and still nothing happens, dump it out and change it for new garbage, a new kind, and more droppings. One hundred percent droppings works, so try that if you can. Try different amounts of water too.

If your container rises but the gas doesn't burn:

▸ Try it for a few days, always checking at night if possible, to see the flame easier. Sometimes we found that the flame is there but weak.

▸ Open it up and add more droppings, stirring them up well.

Our latest version rises consistently 3–5 days in a row with one load of muck. After that it seems to stop producing gas, and that makes sense; any microbes in there ate their fill, died off, and left other microbes that don't produce gas. Real systems have an output port for the old sludge and an input for the new muck. This is just a demo setup.

There is no size limit that I know of to this project. We've made systems three times as big as the one shown here, and on the web you can see family systems in India and Pakistan that are the size of a golf cart. I can visualize one built with various-sized garbage cans. Public systems have been built here in Timor that feed off the manure from hog operations and give a small village methane piped right into their kitchens. The cylindrical slop tanks in these operations are around 5 meters (15 feet) in diameter and 2 meters (7 feet) deep. Given a steady supply of garbage and poop, the sky's the limit!

WHAT'S GOING ON?

We humans often assume we're the supreme organism on this planet, but actually we rely more than you've ever dreamed on microbes—that is, microorganisms, little beasties you can't see without a microscope. You may think of bacteria as bad guys, making us sick and destroying our food. But in our guts there are upward of 10 trillion bacteria, give or take a couple of hundred million, many of which help us digest our food. Along with various other microbes down there, these are known as gastrointestinal microbiota, or gut flora. Such a pompous name for the motley mob of critters.

Some of the stomach problems you could get arise when those bacteria die off. That's why sometimes doctors prescribe yogurt for help with digestion; yogurt has nice bacteria in it, and they can help the gut bacteria factory get going again to process your food.

Not only yogurt but beer, bread, tempeh, stinky tofu, sauerkraut, and some pickles also rely on microbes to ferment these foods in a controlled manner. So some microbes are our friends. The key microbes in your biogas digester are called *methanogens*. They are members of the archaea kingdom, which makes them a bit like bacteria, but not quite. Their normal cycle of life involves avoiding oxygen, eating garbage and poop, and giving off methane. Isn't that so very kind of them?

Some of these guys live in your gut, as well as in the guts of other animals. A lot of them live in the first stomach of ruminant animals, like cattle, that chew their cud and have multiple stomachs. When you or your dog have to pass gas, you may have eaten something that was particularly tasty to those microbes, and they're giving off methane.

Some of those methanogens come out with the droppings, so when we put the droppings into the digester, we're setting up the system for success. If you just use garbage in your digester, you need another source of the methanogens. You can buy a starter kit, take some from a digester already in action, or dip some out of a cesspool, or you can just get the droppings as we did here.

WANTED: BILLIONS OF MICROBES TO WORK FOR FREE

Besides helping us digest food, microbes help us process wastewater. Methanogens and similar microbes are found in septic tanks and wastewater treatment plants that deal with all the poop from an entire city. The microbes take solids out of dirty water and help break down solid waste into products that are easier to dispose of. Some water treatment plants catch the methane and burn it for energy.

Sometimes it's a challenge to get the right microbes growing that will do the job, but other times—as in all the septic tanks I've ever relied on—we humans don't have to do anything at all; the good microbes just move in and grow and flourish with populations in proportion to how much waste is produced.

It's really quite impressive when you think about it. If all those tiny workers suddenly went on strike or decided they no longer wanted to munch on our waste, we'd have serious problems to deal with.

Biogas is usually categorized as clean or green energy, and it's certainly free. But one of the results of burning methane is carbon dioxide, and that's a greenhouse gas. Well, heck, methane is also a greenhouse gas, 20 to 80 times stronger than carbon dioxide (depending on the time frame selected). If anything, we're helping things by burning the methane to make carbon dioxide, if we assume that the methane would have been produced wherever this garbage and droppings got dumped had it not entered our little tinkering project.[2]

The key green aspect here is that families who power their houses with biogas don't have to burn petroleum or chop down forests to meet their energy needs. That's a benefit nobody argues with. And there are numerous side benefits as well: there's no need to get rid of this waste elsewhere, there's no need to go walking to look for other fuel, you can still use the slurry from the bottom of the container to fertilize your garden, and you're not reliant on a public energy system that may not be dependable.

So, long live household biogas! Did you notice the marshmallow tasted just a little bit better than usual knowing that you cooked it with garbage?

ENDNOTES

1. Cattle are actually one of the least efficient sources of calories—that is, the most environmentally consumptive. What's more, cattle produce an inordinate amount of methane, one of the most powerful greenhouse gases, when they burp and pass gas, which they pretty much do all day every day; they can put out more than 50 gallons of methane each day. The only time it really makes sense to eat beef from an environmental perspective is if the land that cow grazed on wasn't good for growing anything else. Dairy products are more efficient when figured over the life of the cow but still not nearly as efficient as plant proteins. That's a real issue for me, for although I'm vegetarian most days, I loves me some cheese.

2. Actually, there are two other main destinations for garbage, here and around the world: gardens or fields and pigs and chickens. Both use garbage in a highly efficient and green manner. We try to use our food garbage to maintain a compost heap to fertilize the dirt in our garden, but several of our neighbors have pigs in a small pen (yes, right here in the middle of the capital city) that love to eat our garbage. We also keep chickens (downtown Dili), and the diversity of their appetites is amazing. Today they cleaned off the fish carcass we had left over from last night, as well as the fruit salad rejects.

Chapter 4

FERMENTATION AND DISTILLATION

Get some fungi to produce alcohol, and then distill the resulting grog.

Timorese make palm wine from the juice that drips from sliced palm branches. It's quite astonishing to find that it's already alcohol as it froths and drips from the gash, high atop a picturesque palm tree. Sometimes certain leaves and barks are placed in the juice to make it continue the process of fermentation and raise the alcohol content.

Like beer, this palm wine is loaded with calories, and it constitutes a significant source of nutrition for many living in the countryside. Sometimes the palm wine is distilled into harder alcohol and sold in used water bottles. Check out the sidebar to see photos of a still I came across traveling in the hills.

MOONSHINE TIMOR STYLE

This is the boiling chamber made from a horizontal barrel with old pots stacked on top of it, forming a tube, all glued together with tree sap and sawdust. A giant bamboo tube emerges out the side of the top of the stack, which is actually an earthen pot with a hole bashed in the side.

Here is the end of the bamboo tube, where the hard liquor comes out. You can see two water bottles already full of stiff drink to the side of the jug being filled.

This is a view from the other end of the bamboo tube. It's two long pieces linked together. You can see other bamboo sections stacked diagonally up near the shack where the boiler is. They are full of the palm wine to be distilled.

Straight, weak palm wine is also available for sale at 50 cents for a 1.5-liter bottle. It smells sweet and fruity.

Two major processes are at work here to be tinkered with: fermentation is biology, and distillation is chemistry. We'll do both, in that order. Yeast is the fungus we'll use to make alcohol. Fungi are life-forms, different from plants in part because they can't do photosynthesis. Yeast eats sugar and produces alcohol and carbon dioxide in the process of digesting the sugar.

Alcohol is rarely made from straight sugar. Wine is made from fruits, in which the sugar is inside and gets eaten by the yeast. But a lot of alcohol is made from grains, potatoes, and other starches. Yeast don't eat starch, but if you let a certain enzyme called *amylase* degrade the carbohydrate molecules making up the starch, you get a bunch of sugar molecules for the yeast to eat. That's what happens in making beer and vodka, among other things.

Where do you get amylase? It just happens to be produced in the seeds of many grains that are in the process of sprouting. This is called the *malt* of the grain. So when you hear of malt liquor you know it's had sprouts in the process.

We use cassava here and also potato and sweet potato to make the alcohol for this demo. Unlike potatoes, sweet potatoes and cassava seem to have enough amylase in them that you don't need to add any more malt. I'll show you two ways to mix up some mash to dump yeast into and get some fermentation going. Your goal is just to see the process—I don't want you going into the bootleg moonshine business!

Gather stuff

- 4 medium-sized sweet potatoes
- Water
- Sugar
- Yeast
- Two large plastic bottles
- Funnel
- Two balloons

Gather tools

- Pot with lid
- Potato masher or spoon, wood spatula, bottle; anything to crush them up.
- Stove

TINKER

We'll make a sweet potato mash first. Chop up around four mid-sized sweet potatoes—we used two different kinds; oooh, look at those purple ones. Mix them with some water and stick them in the blender if you have one. If not, no problem, but you're going to have to beat on them more later.

Put them on the stove at super low heat. The trick is not to let them get above around 150° Fahrenheit. Above that temperature, the amylase falls apart and can't do its job of turning the carbohydrates into sugar. So if you've got a thermometer, keep it right on the money. If you don't have one, just don't let it get anywhere close to boiling. Our stove's low setting was not low enough, so we had to jack up the pot with the other burner stand.

Make sure it's cool—hot water may kill the yeast—and then dump a spoonful or two of yeast in there. Put a balloon onto the bottle and sit it in a nice warm place to ferment. Wait 4–6 days. Those yeast fungi are intrepid, but they're not fast.

While you're waiting for your sweet potato mash to cool or ferment, make a straight sugar one. This one is quick and easy. Fill another bottle most of the way up with water, dump in around a cup of sugar and a spoonful of yeast, and stir it all up. Fasten on a balloon and put it with the other bottle. Watch these bottles over 4–6 days to see what happens to the balloons.

Keep it there for a couple of hours if you can. If you weren't able to blend them, at the end of the two hours smash them to bits with a potato masher or something similar.

Now let it all sit and cool down, maybe over night or even a whole 24 hours. Let those amy-lase enzymes go at it. If you dare, have a taste—any sweetness? That's what you're looking for.

Once this is over and your sweet potato mash is loaded with sugar, pour this all into a 1.5- or 2-liter bottle, and add water until the bottle is almost full.

If you see the balloons fill up, that's a great sign. Remember, carbon dioxide is a byproduct of fermentation with yeast. Actually the balloon doesn't have to get very large. The photos here show one of ours at maximum inflation—not very big—but we still got good alcohol content from both of them.

When you think it's done, take the balloon off and take a whiff. You should scent that nasty beer smell you can find at college parties. The professionals use a thin glass instrument called a hydrometer to measure how much alcohol is in the mash. You can also try to burn a splash of this in a spoon or metal bowl. If it burns, you've done something incredible—there should be less than 10 percent alcohol in there, which should never burn.

WHAT'S GOING ON?

The little yeast beasties are eating the sugar and making alcohol and CO_2, that's what's going on! The bottle with the sugar should have a faster, farther rise of its balloon. That yeast in there is living it up, eating the sugar like candy (I guess it sort of *is* candy), and the results are obvious. At some point they'll eat it all and the process will stop, but the alcohol will stay in the water if you keep it covered. That yeast digestion process is anaerobic, meaning it doesn't need oxygen. In fact, if you give it oxygen other reactions might happen and you might not get much alcohol.

Most people making alcohol boil the potatoes at some point. This opens the cells and tissue of the potato so that the amylase can do its work. But if you boil them first, the amylase will be destroyed and you have to add more. The way I've described works for us without adding additional amylase. Now let's try to distill it!

Gather more stuff

▶ Small pot

▶ Small metal or glass cup or bowl

▶ Two metal tubes such as steel conduit, around ½-inch diameter, 30 and 60 cm long

▶ 90-degree steel pipe elbow joint, ¾ or 1 inch

▶ Chunk of wood

▶ Two large cans of food, unopened

▶ Packing tape

▶ Cookie sheet, depending on the layout of your stove

Gather more tools

▶ Scissors

▶ Knife or box cutter

▶ Stove

▶ Lighter, with a long neck if possible

TINKER ON

To distill your firewater, you need a still! The basics of a still are a boiling chamber, a vertical passage for the vapor, and another longer sloping passage in which the condensation occurs. Here is one we use with the students. It's powered by a small electric burner. The mash is boiling in the black pot and the condensing tube is a length of bamboo. The flaming alcohol shows success—that means the alcohol content is up around 50 percent.

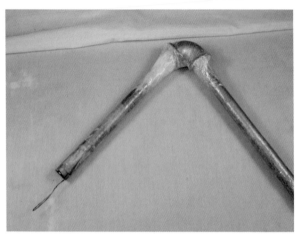

I developed a smaller, stovetop version that doesn't involve bamboo. Feast your eyes:

Make a stand something like ours with two cans (black pitted olives from Spain) and a sturdy chunk of wood. Tape the tube onto the wood chunk right at the center of gravity and the whole thing is pretty stable.

We hit on the idea of using conduit, cheap metal tubing available everywhere, and linking it with a standard 90-degree pipe joint of a larger dimension. Here you see ½-inch conduit joined with a 1-inch joint. This makes it possible to flex it to less than 90 degrees, which is what you want. Stick the tubes in, one on each side, bend them to the approximate angle, and then tape it all shut with packing tape.

Tape a partially unbent paper clip onto the end of the vertical tube so it can rest on the bottom of the pan and hold the end of the tube out of the liquid.

If you have a pot lid that has a handle you can take off and it leaves a large hole, you can use that. Or get an aluminum pie tin that exactly covers your pot, poke a hole in the top that just fits the conduit, and tape it all up tightly.

Set this up on the stove, and before you fasten the lid down, dump the contents of one of the bottles into the pot.

Now tape around the lid so that no air can get out. Don't get the tape anywhere down near the bottom of the pot, or it will melt on.[1] Put the small metal or glass bowl under the output. We had to use a cookie sheet here to extend the

stove surface so the bowl would sit nicely. Turn the stove on low, then sit back to watch.

When you first start to see a few drips in the cup, you can shut the stove off. We want to prove that we've distilled the alcohol out of the mash, so you can test it in two ways. First, you can stick in a finger and wet your whistle. It should sting your tongue a bit, maybe feel a bit warm. Think of St. Bernard dogs bringing gin to victims buried in the snow; it doesn't help them actually warm their bodies but that sensation of warmth must feel mighty nice.

Or you can put a match to it. Move the bowl far away from the output tube and try lighting the drops that splashed in there already. You should get a nice clean blue flame, no smoke. Burn, baby, burn!

TROUBLESHOOTING

If this thing doesn't work the first time, join the crowd. After seeing my chemistry buddy do this with a big bamboo tube and a kerosene stove, I ran home to try my mini-version. About five tries later I finally got it to work. Here are some things that may be going wrong if you're not getting hard grog:

▶   The mash is not boiling. Turn up the heat a smidgen.

▶   You smell alcohol and hear the mash boiling, but nothing is coming down the tubes.

You've got a leak. Go all around all the joints with tape again.

▶   The mash is boiling, but the vapor is not condensing; you can see steam coming from the outlet tube. This means the pipes are too hot. Try putting a wet rag on the sloping one, like this:

The pipe rising from the stove should be pretty hot—you don't want the vapor condensing inside that one or it will just run back down into the boiling mash—and the sloping one should be pretty cold. Our goal is just a few drops of alcohol, so you don't have to worry about the challenge of keeping that sloping pipe cool in the long term.

▶   The first drips don't light on fire, nor do they sting your tongue or taste anything like alcohol. This means your fermentation didn't work, and there is no alcohol in your mash. Go back and make a new mash; make sure it's smashed to bits, and maybe try boiling it after you leave it sitting overnight to release all the sugars that the amylase hopefully created. Try, try again.

If you try all this a couple of times and still get nothing, and you suspect the problem is in your fermentation process, you can try distilling alcohol that someone else fermented. Wine is often 10–20 percent alcohol and beer is 5–10 percent alcohol. Neither of those should burn, so get some, put it into your still, and see if you can burn what comes out. I got some readily burning alcohol drops when I distilled the cheap Indonesian wine and beer for sale here.

In a fit of extreme simplification, I decided that I could actually distill the beer in its own can. I dumped half the beer out of the can, cut a balloon halfway up its neck, fastened it securely over the top of the can, poked a small hole in the balloon, and linked up three straws for the condensation tube. The straws got pretty hot when the steam started through them, so I had to support them with a jar of local peanut butter and drape wet napkins over them to encourage the condensation. I had a few drops of hard hooch in a matter of minutes.

WHAT'S GOING ON?

The theory here is that the mixture of water and alcohol in the pot will be separated due to the two substances' difference in boiling temperature. From a simplistic perspective, the alcohol will boil first, sending its vapor off and down the tube. In the tube, the temperature drops again and this vapor condenses on the inside walls to drip down into the waiting cup.

That's what happens, but what also happens is that some water vapor goes up with the alcohol vapor. Basically all liquids are evaporating all the time. This means you can never get 100 percent alcohol from a still. You can get closer by distilling what you've already distilled, but you'll never get it perfectly pure.

If you leave the stove on longer and try to get a whole cup of the hard stuff, two things will happen. First, the sloping tube will get hotter and hotter and the vapor will be less and less likely to condense in there. The vapor will make its way out the bottom of the tube and waft away wasted into the room. Also, the more you boil, the more water will boil, so what you do get dripping into your cup will be more and more water, and less and less concentrated alcohol.

These first drops are highly valued here in Timor. They are called "the firsts," and they cost more precisely because they are of a higher concentration of alcohol. There is a serious problem with this, though. In many types of alcohol, especially those made from grain, a type of alcohol called methanol boils off first. It has a lower boiling temperature than the one people like to drink—ethanol—and also happens to be a nasty poison. You can go blind from it. But the U.S. Environmental Protection Agency says a few milligrams are not harmful, and that's the most you'd ever get from this finger-licking experiment.

TO DRINK OR NOT TO DRINK

I know a handful of people who are alcoholics, some doing well in controlling it, others in a sad spiral, taking their friends and families and careers down with them into the grim pit of misery that results from this unfortunate condition. A friend once told me, in a dreary, matter-of-fact way, "I'm drunk, but I don't want to be."

The same as with other drugs, and even drugs the body creates for itself—endorphins and adrenaline, for example—if your body likes what it feels, it's going to send you looking for more, even if it is not healthy for you in the long run. It's important to realize that each body will respond to alcohol in a different way; you don't have any choice at all about this response. Many people can drink regularly and moderately and never have an issue. At the same time, many alcoholics have reported that they were addicted from the very first drink they had and spent the rest of their lives struggling to manage this situation, with or without success.

So if you're a kid, think long and hard about this. My grandpa's father was an alcoholic and made life difficult for the whole family. There is a hereditary link to alcoholism. I decided I wanted no part of that, so I've steered clear of the stuff all my life. It's great fun to make, though, and learn biology and chemistry!

Our goal here was to increase the concentration of alcohol through selective distillation. This is an incredibly useful technique and, in fact, is used to produce many things you use, including perfume and gasoline. To be sure it worked, we tasted the result and compared it with the stuff we put in the pot, and we tried to burn it. When the concentration of a mixture of alcohol (ethanol) gets up around 50 percent, it will burn. Again, it's more complicated than that. The alcohol liquid gives off more vapor if it's hot, and the stuff you just made comes out pretty hot, so it may be less than 50 percent and still burn. As mentioned earlier, a more accurate way to determine the concentration of alcohol is with a hydrometer, with which you can check the density.

The point is, what comes out the tube is much more concentrated than the liquid in the pot. That's what distillation is all about. By the way, you could use this same setup to get the salt out of saltwater. That's also distillation, but the goal there would be to get the water away from the salt—that is, a liquid away from the solid that was dissolved in it. That's an easier task than trying to separate one liquid (ethanol) that's dissolved in another one (water), which you just did!

ENDNOTES

1. It's amazing how heat resistant packing tape is—the stuff up near the rim of the pot never melted in over a dozen distillations. Of course, we never let it go very long; as soon as we've got a spoonful of the hard stuff, we shut the still down.

Part II
The Math of Life

Chapter 5

BASKETRY IN TIMOR-LESTE

Open your eyes to another world of containers.

If anthropologists were to count the top 10 critical developments of early technology, basketry would certainly make the list. Imagine living in a world without plastic or metal and having to carry a bunch of stuff, especially stuff that comes in small pieces like seeds, nuts, or little fish. You could carve out a wooden bucket or use a piece of gourd, or sew shut an animal skin, but the best way to make a large, light vessel would be to weave it from natural fibers.

Making a vessel from the leaves of a tree or a handful of grass is an extraordinary accomplishment, one that we modern hominids would have a tough time imitating. Like metallurgy and forestry management, it's an essential skill that has passed from general knowledge to high specialization. And unlike those two, the technology of basket weaving has also passed into the endangered category, at least in the good old USA. I can imagine there are massive regions of our country where not a single person knows how to weave any sort of basket.

And why should they? Three dollars and ninety-five cents gets you a spiffy little plastic bucket or laundry bin, straight from a factory in Asia. Life is short! Why learn a useless skill?

I'll tell you why, you little whippersnapper! Baskets made from natural fibers have several key advantages over plastic and metal; they're lighter, they're more beautiful, they feel nicer to the touch, you can make them yourself for free and get a little charge of self-satisfaction, and maybe most important, the materials they're made from are entirely renewable resources and biodegrade easily back into the earth. Learning to weave baskets also gives you insight into our ancient history, a taste of the very essence of our development as a species, and respect for what traditional peoples still do today to survive with minimal resources in a highly sustainable manner.

In these chapters on Timor's basketry, I'll show you some of the magnificent mathematical discoveries we've made in the realm of basket weaving here in Timor-Leste. I've gotten spine-tingling satisfaction from learning this stuff, all of which came to me, directly or indirectly, from mostly uneducated Timorese women, often from the rural hinterlands of Timor.

GRANDMAS IN TIMOR-LESTE

I did a TedX talk a few years back[1] trumpeting Timorese grandmas' pedagogy in mathematics and science. I demonstrated various highly technical concepts that any grandma must have known to, say, get the basket to come out right, separate the chaff from rice grain grains, extract salt from tidal soil, or liberate the oil from a coconut. She has taught these concepts as she has learned these concepts, completely removed from any sort of school or education bureaucracy. I believe we could gain a lot from studying her highly specific methods.

I've been continually astonished by the amount of science and mathematics that the old folks around here have learned informally to carry out the sorts of day-to-day, traditional, life-sustaining work they do.

Now, I'm not saying these elders can describe these concepts in a coherent academic fashion, much less use standard terminology to articulate what they do. But this is the realm of what has been described as "submerged" mathematics or science. It's under the surface and not often seen or remarked upon, but the technical content is absolutely undeniable. From the perspective of people who rely on baskets or salt or coconut oil, the formal articulation of the phenomena at play is not a high priority. The priority is the end result, and the phenomena need be understood only to the extent that

they can be manipulated toward achieving this utilitarian goal.

To a person, each of the elders I've talked to has expressed interest in hearing the formal articulation of these concepts. Through the national curricula my colleagues and I have developed around ethnoscience and ethnomathematics, their grandchildren now get the opportunity to learn about this stuff in schools.

I was made to learn an extraordinary amount of mathematics to get my physics degree, yet even as I write this book in the process of what has been 10 years of on-and-off research into basket weaving in Timor, I continue to learn more mathematics from Timor's women weavers. So follow along for a bit, and see if you get the same rush I do.

Let's look at a few major forms of basket weaving in Timor-Leste. Take a look around at what all Timorese live with on a day-to-day basis.

Bote: The Bucket Basket

The name for this basket and for a bucket is the same in the Tetun. It's woven with leaves from the sago palm, which are strong strings. You heard me right; it's not that they are used to make string, but rather that you can rip them off the branch, twist them a few times, and use them to tie the tarp down on your truck, or tie the roof onto your house. Twist several together and you can tie up a two-ton water buffalo. A group of students and I did a test to find the tensile strength of one of these *bote* baskets, and when it was stuffed full of rocks and all the pieces of steel we could find, it had not broken yet, so we called off the test.

You can see it's a sort of standard weave, sometimes called *twill,* the strips going over two, under two. But check out the bottom; it's based around an entire chunk of the palm branch. Here you can see three coming together at the base, with the strips coming out being woven tightly around and up the sides.

Check out the strap. It's woven with the same sago palm leaves, broadens at the place where it loops your shoulder or forehead, and gets integrally incorporated into the weave of the basket so that the *bote* is supported bottom to top and all around the rim. That's engineering, my friend.

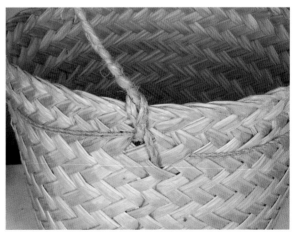

Lafatik: The Winnowing Basket

This basket is in every house in Timor, without exception; everything stops if the *lafatik* is missing. It is used to clean rice and other foods before eating. When we first came to Timor, we bought some local rice and, upon getting ready to cook it, realized it was still full of hulls, unhulled grains, and other bits of crud. Washing the rice did no good; the unhulled grains and half the hulls stayed together with clean grains. We picked the hulls out one by one with our stubby fingers, feeling entirely inadequate for the better part of an hour before we had enough edible rice for our meal.

Later we learned that you need to put the rice in the *lafatik*, then fling it up into the air to separate the hulls and chaff. This is called *winnowing*, and every grain harvest for the last six thousand years has involved it in some form or another.

Winnowing works because the grains are denser with less surface area, so they have less resistance and fall faster through the air. The hulls and chaff are less dense with higher surface area, so they fall slower through the air—more resistance. It's all based securely on physics, but it's also quite an art. In every household here, someone knows how to do this: fling it up, catch the falling grains,

and let the hulls fall on the ground. Here's our friend Mana (Sister) Joana demonstrating.

Now, the unhulled grains are still there! They're approximately the same density and surface area as the hull-free grains (the ones we want to eat). Enter again: the *lafatik*. Now it's a rotating and shaking motion that moves the grains around in the basket; the unhulled ones come together as if by magic in the center, where they can easily be skimmed off. I've heard this called the "Brazil nut effect" in some circles; you may notice in other situations that bigger things come to the top of an agitated mass of particles. Again, the master hands of Mana Joana. See the hulls congregating in the center?

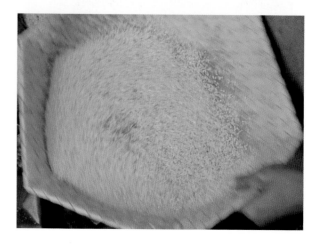

So this basket is a life saver, because with those two motions, the *lafatik* makes rice from the field ready to eat.[2]

This hexagonal beauty is one of our richest sources of mathematics. It's a weave we call the rhombus weave because it's made entirely of small rhombuses—slanted squares, or diamonds—which can in turn be decomposed into two equilateral triangles.

Three of these rhombuses make a hexagon, and six of them make a star.

Notice the angles in this weave are completely different from the *bote* weave earlier and the *kohe* weave later. More on that later.

Mamafatin: Betel Nut Basket

This cute little bugger is used to store your betel nut, betel leaf, and lime powder, and especially to offer it to valued guests, such as your future in-laws. Chewing these three things together is a common way to get a mild head-buzz here, as in many parts of South and Southeast Asia. Then you spit the disgusting red juice on the floor, wall, gutter, steps, etc. It can smell pretty good, I think, a bit like menthol, but in general it messes up your teeth and puts you at risk of lip and gum cancer. Nonetheless, when the in-laws show up here in Timor-Leste, you'd better have some of these three things to offer, and it's got to be in a fancy little *mamafatin* basket.

Notice how the lid is another shallower basket, seated into the bottom section—clever, eh? Various baskets in Timor have this top that fits tightly into the bottom, and I have to think it's quite a challenge to get that to happen as you weave.

The *mamafatin* is more rhombus weave, like the *lafatik*, this one more complex, with points and curves of various angles. Note the pointy little feet, with sharp angles, all created with rhombuses. (This one has a bit of drawing on it from previous analysis.)

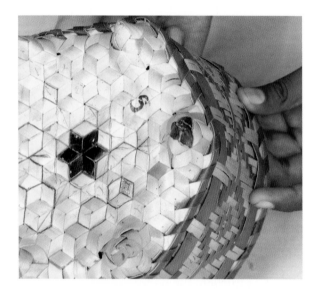

Kohe: Multipurpose Carry-all

One of the most popular folk songs here says, "I left the house with just a piece of cassava in my *kohe,* to hold my belly tightly during the heat of the day..." The *kohe* is like a purse to women and a duffle bag for men. It's woven with sago palm leaves with the strap built into it like the *bote,* but the strips are narrower and finer, and they're formed into a simple square weave. It has a characteristic square bottom transitioning to round sides.

The *kohe* can be made arbitrarily large. Here is one around a meter tall, holding about a ton of rice, straight from the field.

We'll make a little *kohe* as the culminating weaving activity.

CREDIT: Luis Nivio Soares

Katupa: Rice Dumplings

Finally, I'll show you one of the most commonly woven objects in Timor-Leste: the *katupa*, a palm-of-your-hand-sized vessel into which uncooked rice is loaded with coconut milk and turmeric to be steamed or boiled into dumplings. When it's cooked, the palm leaves are removed and the dumpling inside is eaten. *Katupa* are woven by the hundreds for festival meals, and they enable you to conveniently take a rice-based meal on the road. Most people know two or three designs, and I've seen dozens of different ones.

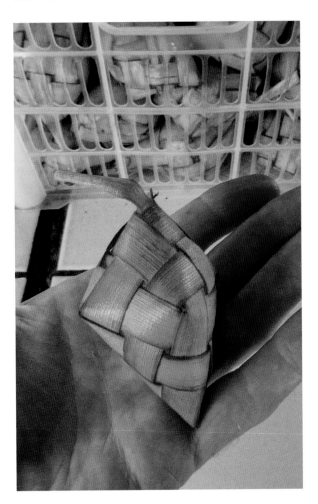

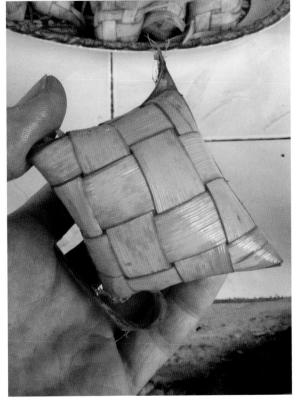

Find more info on these fascinating three-dimensional geometric solids in Chapter 9.

Now that you've got a glimpse of the basketry riches of Timor-Leste, let's check out some of the math hidden within.

GATHERING THE LEAVES

Weaving is hard enough when you start with the palm leaf strips ready-made. It's an additional chore to prepare the strips. I stopped on the road to buy a couple of sago palm fronds to use for weaving.

Here's a girl going up a ladder near a young sago palm tree to get me one of the leaves that's just emerging. Called *dikin*, these are the ones best for weaving. The bottom branches have already been harvested. It's hard to see but there are rows of sharp spikes along the edge of the branches like a shark's jaw. That makes for good fencing material, if you can get it set up without tearing your hands open. She's got a machete in her right hand with which she deftly whacked out the *dikin*. You can see it falling back over her shoulder to me and her cousins waiting below.

When they had chopped a few *dikin* and fronds for me, I asked the girls if they knew how to weave, and then harangued them when they said no. Their grandma was standing nearby, so I told them to go learn something from her right now. She heard us and quietly went over to one of the larger fronds the girls had gathered for me and began weaving it into a shade for a porch or an animal pen.

She stood up when she was done and said, "Yeah, these girls are lazy butts. They don't want to learn anything." I told them one more time to listen to their grandma, took my leaves, and went home.

PREPARING THE STRIPS

Mana Joana helped me prepare the strips once I got the leaves home. Here she's holding up a *dikin* that we've just cracked open. The yellow color means it will be good and flexible—perfect for weaving.

Then each strip has to be trimmed to a uniform width. This is done with a little device fabricated from one of the spines we just chopped out. Mana Joana cut out the spine, bent it twice, slit it in the middle, inserted it into itself, and pressed the knife in just the right spot. Each leaf was passed through in a smooth, mechanized operation, and they all came out exactly the same width.

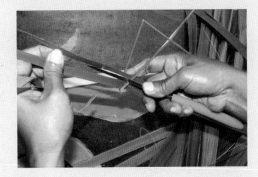

There are about 50 of the double-sided leaves in a frond, all of slightly varying length. We laid it aside to dry for a few days. The drying process shrinks the leaves a bit, so you want that to happen before you start weaving. We went to the open frond and began cutting the two sides of each leaf from its central spine with knives. You can see the skeleton of the frond right in front of Mana Joana and the removed strips piled at the bottom of the photo.

These were dryer than the leaves from the *dikin* frond, so we could start weaving with them right away.

ENDNOTES

1. Search "Tedx Dili Gabrielson Grandma." They missed some of my photos, but you can get the idea. You can get a full PDF of the presentation here: www.dropbox.com/s/m5hwan9zett3b97/Learning%20from%20 grandma.pdf?dl=0.

2. You may say, what the heck? I've never had to worry about this at all! Yeah, well, that's because some enormous high-tech, petroleum-powered, stainless steel machines near the giant mechanized rice fields in central California did all this work *for* you, and the rice was sent off to your supermarket clean in a plastic bag, maybe even with vitamins added. That's nice, when you can get it, which they can't here. Thus, the *lafatik*.

Chapter 6

BASKETS FULL OF MATH

 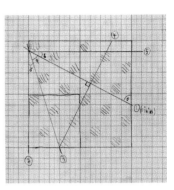

Do serious math with weaving.

Oodles of mathematics await the careful observer of Timor's basketry. I must break it to you that many of the most fascinating secrets we've found in the baskets here will be nearly impossible for me to pass along, because you can't buy these remarkable works of folk engineering at your local big-box store, so you won't be able to hold them in your hand and discover these secrets for yourself.

For example, exploring the mysteries of the hyperbolic surfaces, counting surface areas in terms of woven squares and then comparing that to the formula result, discovering the factors and divisors in the pattern around the walls of the betel nut basket, or using trigonometry to figure the angle of the strange slant of the *raga* weave (that photo up top here): these things are just not going to happen in this here book.

Never fear: I can still show you a bunch of interesting weaving math to tinker with.

Gather stuff

▸ Something with a square grid that you can mark on: mat, basket, checker board, or graph paper

▸ Optional: Something with the rhombus pattern (which I explain in a bit)

Gather tools

▸ Ruler

▸ Protractor

▸ Pencil

▸ Scientific calculator

TINKER

Check this out: In anything woven, you find the surface is composed of small figures; on the local mats, they're squares. The surface can also be seen as a grid of lines and intersections between these figures. If you connect the intersection points, you'll get lines going at many angles.

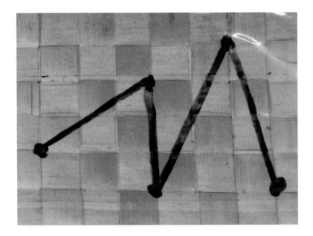

(I've put clear packing tape over the mat, so I can use dry erase markers on it like a whiteboard. My clever Timorese colleagues thought of that one—it's worth a Nobel Prize for teaching.)

Some are special, like the 90- and 45-degree angles.

On baskets with the rhombus weave, you can find 60- and 120-degree angles all over the place. These two angles are based on that tiny, ever-present rhombus itself.

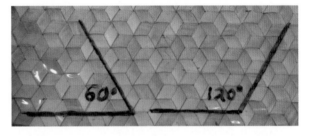

This fan is made with the rhombus weave, and all the edge corners are 120 degrees, either the inner or outer angle.

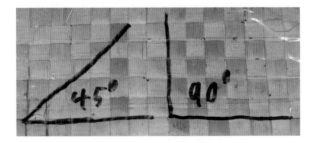

These two angles come up all the time in the square weave baskets and the mat. Look at the edge of the mat; it's all 45-degree angles where the strips turn around to get woven back into the mat. And of course the corner is 90 degrees.

Other angles between arbitrary lines are harder to figure.

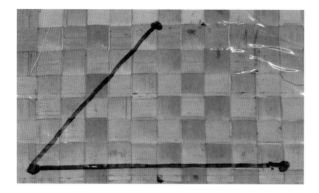

MEASURING AND CALCULATING ANGLES

You can measure these angles with a protractor. If you don't have one, please put the book down and go buy one. Like a ruler or tape measure, it's one of the essential tools of a mathematician, as well as a physicist or an engineer. Here's how you use it: Put the line at the bottom of the protractor on top of the line leading to the vertex of your angle. Put the center dot of the protractor right on the vertex. (If you're reading this on a device, zoom in until the image fills your screen and slap your protractor right on the screen.)

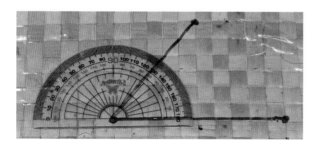

Now read the number where the other line crosses the protractor going up. Careful; there are usually two numbers. On the protractor shown here, the inner one is red and the outer

one is black. One is the resulting angle if you count from the left end of the protractor up, the other the number if you start from the right end and count up. The two numbers added together will always equal 180 degrees, which is a "straight" angle. Notice the two zeros, one on each side of the protractor?

Which one of the scales do you use? You'll always know, because you'll start from the zero at the first line and count up to the other. In this picture, you'll start with the red zero on the right at the bottom and count up to around 52 degrees.

When you make one line coming up off of another line, you'll get two angles, both of which you can measure with your protractor. I'll call them α and β, the first two Greek letters. Mathematicians often use Greek letters for angles in figures, to keep them separate from lengths.

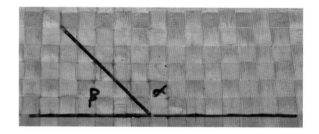

Put your protractor smack on that center point, and you can measure both angles without flinching.

See it? The right side one, α, is 135 degrees measured on the red scale, and the left side one, β, is 45 degrees, measured on the black scale.

You can make and measure all manner of angles on a mat like this; go for it! Draw more angles and measure them with your protractor!

You can also calculate angles using the tangent function and a calculator. Never heard of the tangent function? Follow me: First I'll draw a horizontal base line. Then I'll connect the opposing corners of three squares stacked up vertically, then continue the line upward until it's long enough to cross the protractor. Then I'll form a right triangle with a vertical line back down to the base.

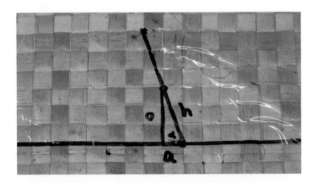

Here I'll call the inside angle at the lower right alpha—α. The sides of the triangle are given names in relation to that angle. The left side vertical one here is called the "opposite" side—o—because it's sort of opposite, or in front of, the angle. The short one at the bottom is called "adjacent"—a—meaning it's right beside the angle. The name of the long diagonal one labeled h on the right you have probably heard before: "hypotenuse."

The tangent of α is defined to be the length of the opposite side divided by the length of the adjacent side:

$$\tan \alpha = \frac{Opposite}{Adjacent} = \frac{3}{1} = 3$$

You can look up the answer for what angle has a tangent of 3. In the past you had to look it up in tables in the back of your math textbook, but now you "look it up" on your scientific calculator (which is probably waiting for you on your phone or device—take a look). You use the button tan[-1]—"inverse tangent"—which means we know the tangent and we want to know the angle that has that tangent. So push tan[-1] and then push 3, then =.[1] You should get 71.56blahblahblah.[2] We can round that off to 72, and that means the angle α is more or less 72 degrees.

$$\alpha = 72°$$

Ain't that slick? This kind of mathematics is called trigonometry. You usually get it around 10th or 11th grade, but here it is, right in the middle of a palm leaf mat.

Just to confirm this, while 71.56blahblahblah is still on your calculator, push the tan button and see what you get. It should get you back to 3, meaning that's the tangent of this angle.

You can also confirm it with your protractor—there in black you see 72 degrees, right on the money.

Here are some more diagonals of triangles, and you can find the angle α by using the inverse tangent function on your scientific calculator.

Just hit the tan⁻¹, enter the number of squares opposite the angle in question, hit the divided-by button, enter the number of squares adjacent, and then hit the = button. Set your protractor down on the line to check it. Then make some yourself and try them too!

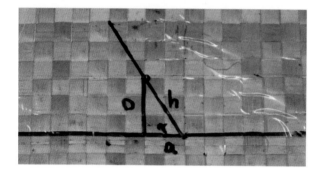

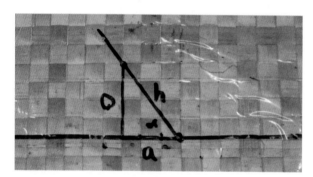

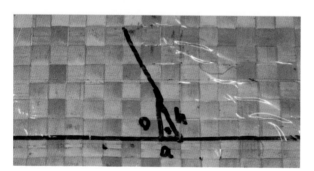

You can do that with any triangle you like.
We used the tangent to find these angles because it was easy to read off the length of the two perpendicular sides. There are two other main functions in trigonometry—the sine and the cosine—and they both involve the hypotenuse, which would be harder to figure on these palm-leaf mat triangles.

GRAPHS AND SPECIAL TRIANGLES

Let me show you a few more things we use these mats for. If you can find any sort of woven object, or anything else with a square grid, you can do these things too!

You can make a bar graph.

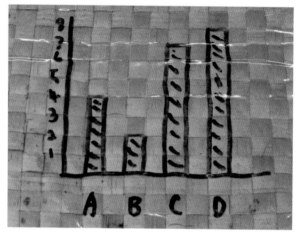

You can make a Cartesian plane and graph points on it.

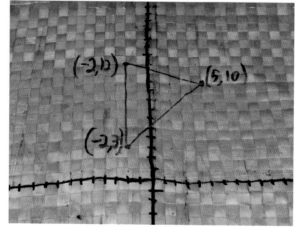

You can show how square numbers and square roots are related.

You can show the relationship to square and triangular numbers.

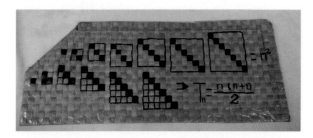

Now let's tinker with some special triangles. Some you've seen already.

Use your protractor to measure the three angles in the following triangles and write down what you get. I'll name the angles alpha, α, beta, β, and gamma, γ, the first three Greek letters.

1.

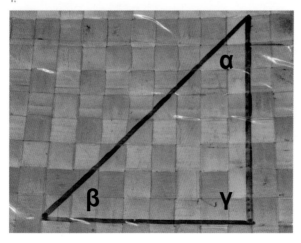

2.

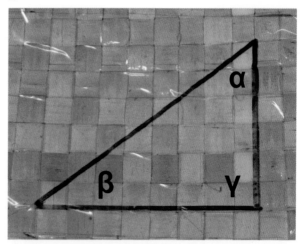

3.

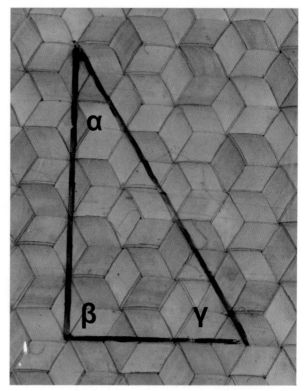

4.

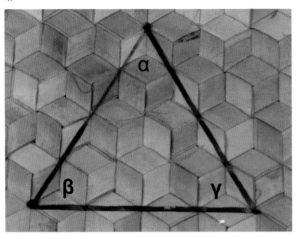

Here's what we got:

	α	β	γ	NAME
Triangle 1	45	45	90	45-45-90 Right Isosceles
Triangle 2	53	37	90	3-4-5
Triangle 3	30	90	60	30-60-90
Triangle 4	60	60	60	Equilateral

Triangles number 1, 3, and 4 are old friends you'll get to know intimately whenever you do geometry. Triangle number 2 is also cool. Its name comes from the length of the three sides. This one is actually 6-8-10, which is twice 3-4-5. That's nice, but the angles don't come out so neatly. (You saw that one already, up in the trigonometry part above with a triangle made in a set of squares 3 wide and 4 high.)

Did you notice that these inside angles always added up to 180 degrees? Good; now you can believe it when your textbook tells you that.

Did you notice triangle 3 was half of triangle 4? Those two 30-degree angles add up nicely to 60 degrees up top.

Notice also that those last two triangles were drawn on the *lafatik*, because of its different weave that creates 60-degree angles all over the place.

But now enough of this mathematics. Let's do some actual weaving!

ENDNOTES

1. On some calculators it's the opposite: first push 3 and then push tan⁻¹. Fiddle around until you figure it out.

2. Don't worry about those numbers that seem to go off to infinity. They do, and they don't matter much. Always be ready to round off and ignore them; a "more or less" answer is good enough when you're tinkering. It's nearly always a mistake to write them all down; it's like counting the distance up your first front step when calculating the distance from your house to the Statue of Liberty.

Chapter 7

THE RHOMBUS WEAVE

Form simple shapes with one key pattern.

Now that you've had an overview of the math in basketry here in Timor, let's weave something. We can make simple shapes and even letters with the rhombus weave on just two strips. It's only for decoration, but you'll get the gist of the sort of weave used in the *lafatik* and *mama-fatin*, among many others.

Gather stuff

▸ Several long strips of paper with uniform width

GETTING YOUR STRIPS

If you don't have palm trees around, you can use reeds or grass. Go see what is growing in terms of long, slender plants in your local wild area.

You can also cut strips of any paper and weave them. Stiffer paper is nice, and file folders work well; you should be able to nab some of these from your nearest office. You can just use scissors, but see if your local copy shop or your school has a big old paper cutter to get nice, uniform strips. Fold the folder normally so that when you cut a strip, it opens out into twice the length.

If you want really long strips, try using paper rolls made for adding machines. But then you're limited to white, and the paper is usually a bit thin. Thin strips are available by the bazillion in paper shredders, but they tend to be short and a bit hard to work with.

A way to get nice colors and strips of around one meter is to get a piece of colored poster paper—moderately stiff, not the really thick stuff—roll it up, and chop off rings from the end. After you roll it, tape it off and then mark the width of your strips. We used 2 centimeters.

Now get some hefty scissors, squash the tube a bit, and cut out the coil rings.

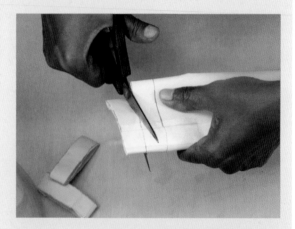

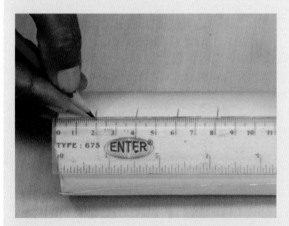

Make more marks around the tube and connect them. It works well to restrict the tube with fat books or rocks or something, and then hold your pencil steady while you rotate the tube.

Finally, take off the tape, unwind the coil rings, and roll them up again gently in the opposite direction, to take the curl out of them.

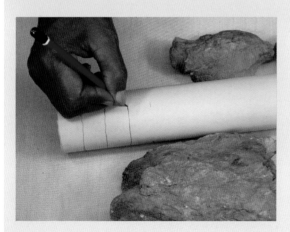

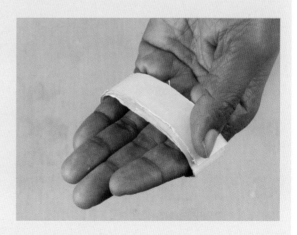

Now you're ready to weave!

TINKER

Start with two strips, one (horizontal) on top of the other (diagonal). Here is the key to this type of weaving: the 60-degree fold. You can see the angle made by the pink strip going up in front and back down in back, with three edges meeting at the point. Getting all three surfaces to come together at a single point ensures the correct angle.

Now fold one end of the blue horizontal strip over to run parallel to the pink strip.

And finally, you can weave the pink one on one side over the blue and under the other pink.

Now turn the whole thing over.

Wow, it looks exactly the same! But don't be thrown off. The one you want is the blue on the far left in this photo. Weave it up over the pink and under the other blue.

Now you can see again the two parallel strips in the center. These are the ones that will be stationary—the base strips—and the two angling off to either side will be woven in and out of the two base strips. Continue with that blue strip you just used and go back up over the blue base strip, then under the pink one. Always leave the strip perfectly parallel to another one when you lay it down.

Flip it over again and grab the pink. Do two moves: over and then back.

You get the picture: over-under-over-under and flip, alternating strips. Always try for tight folds, parallel placement, and no space between the strips. Keep on going until you come to the end of your strips.

Now take one of the non-base ends back across as if you were going to continue going straight.

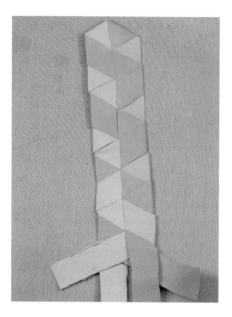

But then take that base pink and weave *it* across to come out on the far side.

That's the base technique for using the rhombus weave on two strips to get a straight line. Of course life gets pretty dull if all you've got are straight lines. There are two main ways to turn corners with this kind of weaving. The first gets you a 120° angle. Start with two more strips just as you just did and weave until you have a short section going straight.

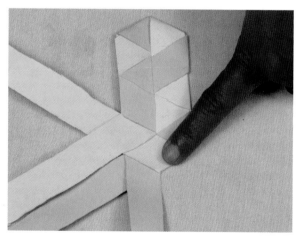

Ooooh, that's new! Now you've changed directions. The two in the center are now your base pair, and you can weave the others back and forth across them as usual. Keep on weaving for a bit.

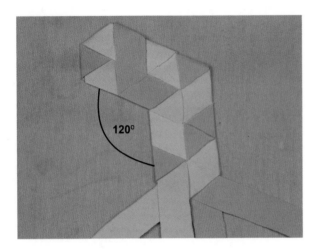

You've made a 120° angle.

The other way to turn makes an even sharper angle. Start with two more strips and a short straight section.

But instead of continuing on straight, take that blue strip on the side, weave it over, and then take the other blue strip and weave it over *again*.

Now turn as you just did, swapping one of the sides for a base.

Now once again the center two will form a new base pair. Keep that clear in your mind and weave the other two on along them. Remember, every two steps you have to flip it over.

You've made a 60° angle.

And there you have it: the three possibilities for progressing with this particular weave. Now if you want to continue on any of these, just staple or tape on another strip to one of these and weave on into the sunset!

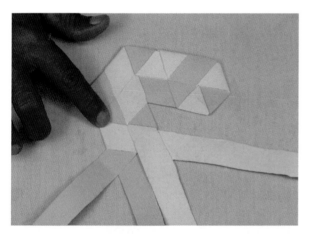

WHAT'S GOING ON?

We made turns at only two angles, each related to that original 60° angle. This is remarkable, because the same is true for enormously elaborate basketry made from this weave. Here are some beauties:

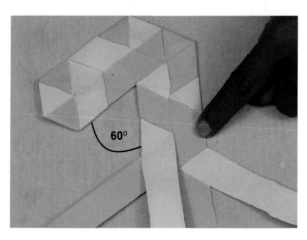

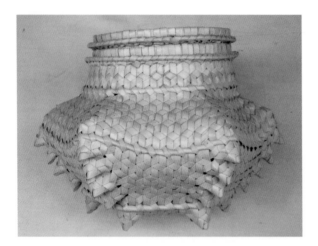

With only these angles you can also weave all sorts of shapes, including letters, numbers, spirals, and boxes. Closing them off is a bit tricky sometimes, but basically you're just looking for how to fold and weave the ends back into the existing piece, and then clip them off. Here are some figures woven by Joana and me and some of my students. Fill your room with them!

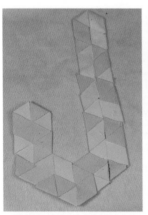
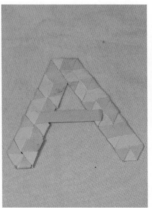

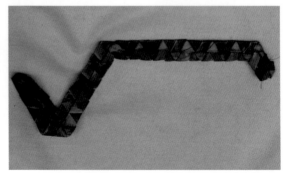

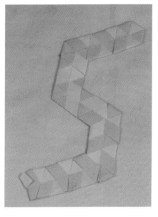
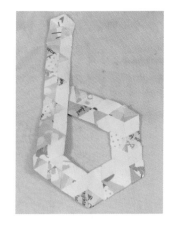

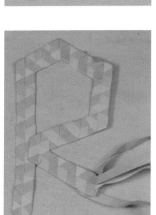
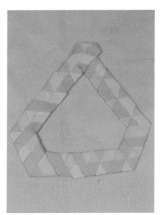

Chapter 8

KOHE

Make a useful little basket with the square weave.

This is a super-basic basket called the *kohe*. Timorese weave it using strips of leaves from the Sago palm. Mana Joana used a *kohe* to put the seed kernels in when she went to plant corn. She recalls the pride of first learning to weave one for herself in late primary school.

Gather stuff

▸ 8 strips or more at least 50 cm long

Gather tools

▸ Stapler

▸ Paperclips, small

▸ Marker, pen, or pencil

TINKER

Start with the under-over weave of the placemat you may have made in kindergarten, four horizontal and four vertical.

Timorese grandmas would never start like this. They'd be clutching the strips in their hands on their laps, introducing them one by one, weaving them in and tying them off so that they're secure. But we've decided this is the easiest way to show the steps on paper.

Now we'll use another trick the local grandmas would never dream of: the stapler. That four-by-four grid needs to stay in place as we make the corners and form the walls, so put staples in four corners and make the grid solid.

Now here's the tricky part. The base of the basket will be a square, but not that four-by-four square you just stapled. It will be centered in that square, with corners at the four center points of the sides. So put dots at those points, and then draw lines between them, just to make it clear where you're heading.

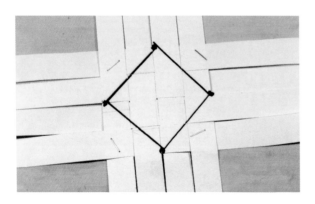

Notice that the staples are all outside that square; they'll be on the sides of the basket.

Now grab two strips coming out of one of the sides, one on each side of one of the points you just drew. Make a corner by twisting them a bit and weaving them together. The ones to their sides will also get woven in.

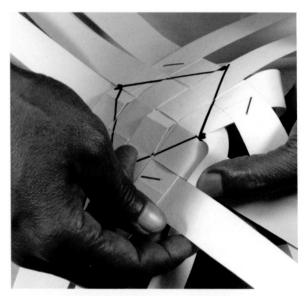

There's your first corner! When you've got to this point, grab a paperclip and stick it on so that it doesn't unravel when you go to the next one.

It's hard to show the next few steps, but basically, all the strips need to meet in that familiar over-under-over fashion, all spiraling around the walls of the basket up toward the top. As you bypass the paperclips, slip them out.

Now go do that same thing to the intersections at the other three points, until they've all formed corners. Your square base should be clear now.

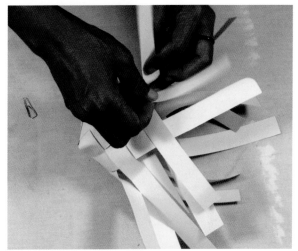

You can keep going as long as you want, stapling on extensions to the strips when they run out. We stopped before our strips ran out, and it made a nice size for marbles or mints.

As you move up, you can tug the strips tight and fold them back to keep everything tight. When you reach the level where you'll make the rim, fold them back all around the basket.

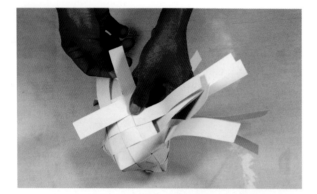

Work out that fold, outward and down, and go all around the basket doing that fold on those tips, ignoring the others for now.

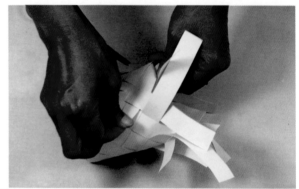

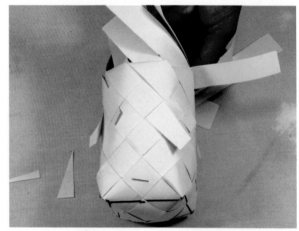

There you can see all the strips in one diagonal direction folded back over their counterparts. Now you'll start doing 45-degree-angle folds and weaving the tips back down into the walls of the basket. To make this easy, you want to trim the tips of the strips so that they are easy to insert—make them a little pointy.

Now do the same thing with the other strips but going the other diagonal direction also from inside to outside.

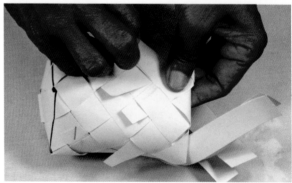

This should form a nice, smooth rim.

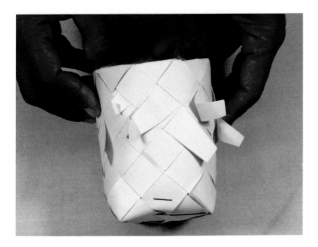

Now trim it up. Some tips you can stuff under the next strip so they are out of sight; others you have to cut off. When you're done, it's beautiful!

You can decorate it, or you can make another one with different-colored strips for a nice, colorful effect.

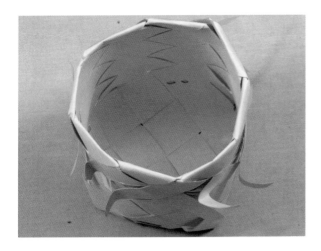

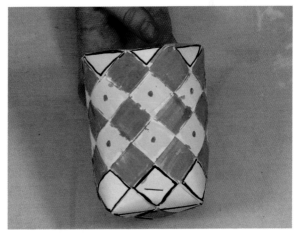

Here's a bigger one, starting with a 6×6 base.

Many times you see *kohe* that are not quite finished like this, bristling with the raw ends of the strips. They get the job done, and life's too short to make everything beautiful, eh?

But then there are these oh-so-beautiful pieces, complete with lids, sold for a couple of bucks.

You can't fully appreciate it without cupping it in your hands, smelling the palm fragrance, and feeling the satisfying grip of the well-fitting lid. It was definitely a professional job.

WHAT'S GOING ON?

Congratulations! You've succeeded in using a two-dimensional strip to form a three-dimensional basket. When your far distant relatives pulled this off, maybe 2 million years ago, give or take a few weeks, they were stoked! Up to then they'd had to use animal skins, animal organs, gourd shells, or hollowed wood to carry stuff. Now a whole new world of packing possibilities was available to them.

Check out those corners you formed: three squares coming together to make a point.

It's the same as the one on the *kohe* itself.

That's the transition from the flat base to the standing sides. Each square has 90-degree corners, and three coming together gives you 270 degrees. But that's not enough to make a flat surface—you need 360 degrees for that. Thus, the basket corner curves and the basket stands up. You'll find corners like that in most baskets, even the little *katupa* woven for rice dumplings.

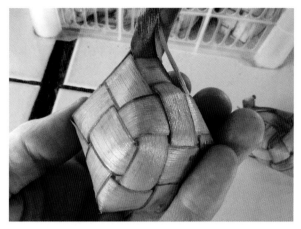

Some corners are pointier than others. If only two squares come together at a point with two sides in common, ordinarily you'll get two squares smashed together. But if you puff up the space between them a bit, you get the tip of the octahedron *katupa*.

 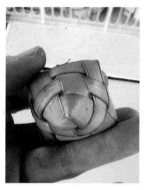

Read more on that and the angles formed when polygons come together in Chapter 9.

The point (ha!) once again is that the mathematics in this weaving is deep and wide. It's also extremely useful. Before plastic or metal, this is what our human race used to carry stuff. And still today, baskets remain a highly sustainable, nonpolluting, fashionable, aesthetically pleasing, do-it-yourself answer for storing your junk and schlepping it around. See how many you can incorporate into your life.

COMPARING EXPERIENCES IN TWO CULTURES

Our friend Mana Joana helped me a lot in figuring out all this weaving. She is my age, and like me when she was young she lived on a farm where she helped her parents work the fields. That's where the similarity ends.

I helped Dad drive our tractors and trucks and maintain the implements—plow, disk, harrow, planter, cultivator, sprayer—and eventually harvest the fields with a giant combine that picks the grain and threshes it off the cob or stock, and then dumps it into a big wagon with a hydraulic lift that smoothly slides the grain out the back hatch and into an elevator, taking it to the top of one of our three towering corrugated steel grain bins. Although we worked like crazy, staggering late and exhausted into our chairs at supper time during harvest season, all those machines were powered by petroleum, and our labor was directed primarily at the machines, not the earth itself.

Mana Joana's family had no machines beyond simple hand tools and the baskets they made by hand from palm leaves. They borrowed her uncle's small herd of water buffalo to "run the rice paddy," loosening muddy soil before planting (a technique I've seen nowhere else in the world). This is the only way they employed animal labor. She said they would have used horses to carry corn back from the field at harvest, but none of their family had horses, so they carried it in baskets on their heads. Push carts would not have helped on the narrow mountain paths. The rice field was closer to the road, so they paid a truck to carry the rice harvest to their house in large baskets they had woven themselves.

My family's farm was far more productive in terms of net output, but also more consumptive and tremendously more polluting. We dumped loads of pesticides on our fields as well as synthetic fertilizers that took even more petroleum to manufacture. What's more, the grain we raised, mostly corn and beans, went entirely to livestock feed, thus reducing its efficiency. It takes much more land to raise a calorie of meat than it does to raise a calorie of a similar source of vegetable protein, like beans.

My family also consumed a heck of a lot more stuff in our day-to-day life—clothes, food, non-water drinks, toys, entertainment, recreation, vehicles, more petroleum—not to mention the infrastructure surrounding us in modest middle-class Missouri: systems of water, natural gas and electricity; roads and bridges; schools, medical facilities, and government agencies of all kinds. That is to say, we enjoyed a higher standard of living.

Were we happier? I've got two answers. First, it's the wrong question! I prefer presenting the question in terms of justice: were we consuming more than our fair share of the world's resources? I've come to the conclusion that we absolutely were. Divide available resources by population and we were, and continue to be, taking too much. Meanwhile, Mana Joana's family didn't have access to adequate medical care, nor did they have food security in the years when war or weather ruined the fruits of the fields. She had electricity, but she had to carry water to her house. Her Indonesian-run school had no significant resources aside from desks, chairs, and chalkboards.

I didn't ask her whether she was happy growing up, or if she'd have wanted some of the things I had. It seemed a silly question. Like, would you prefer an arduous, insecure, at times unhealthy existence, or a stable, secure, and less tiresome existence? Like, would you prefer crawling down the road or riding in a car?

One time she did tell me of the worst time in her life. It was during middle school and some of the crops failed, so they were without normal food for a season. They resorted to eating sago, a starchy substance obtained from the trunk of the sago palm. Various groups in Southeast Asia rely on this as their primary source of calories, but her family was not used to eating the bland substance and it is extremely laborious to produce: chopping down the tree, hacking fibers out of the fallen trunk, pounding them until they separated from the starchy pulp, then employing multiple filterings to remove the non-edible bits. Mana Joana's family was working the fields hoping for a better harvest next season, and she was left to sweat and cry day after day as she extracted the sago by herself, then cooked it for supper. I'd say she was less than happy.

Here's my second answer to the question: Heck, yes, we were happier, from the perspective of human rights. Aside from the difficulties of meeting life's basic needs, Mana Joana had to deal with the Indonesian military abusing the people of her community and having her life controlled by an undemocratic foreign government. That'll make you unhappy any day of the week. As I've explained in the appendix, as a U.S. citizen, I feel an acute responsibility for this and see my work here as a sort of teeny-tiny reparation effort.

At the same time, though I'm here to give, nine times out of ten I end up receiving more than I've given. I've learned books-worth of knowledge from Mana Joana about life and living here in Timor, shiny gems of wisdom tying the traditional past with the uncertain future of myself, my family, and all of humanity. I intend to go on learning from her as long as I can. Thanks, Mana!

CREDIT: Bryce Johnson

Chapter 9

STICK SOLIDS

Construct geometric solids with tooth-picks and bits of flip-flops.

Most Timorese women know how to weave several *katupa* designs. (See Chapter 5 to find out about the wonders of the *katupa*.) I know how to weave three. They took me hours to learn, and many more hours to relearn after I subsequently forgot them, again and again, as time passed and I didn't make them on a regular basis.

Katupa are also woven in the miniature with dyed strips and hung around the house as colorful decorations. The one at the top here is in the shape of a bird.

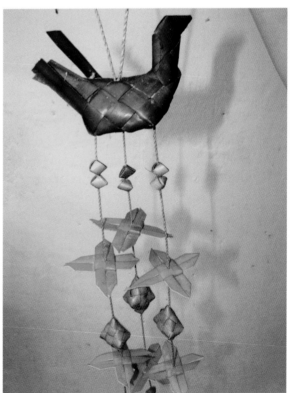

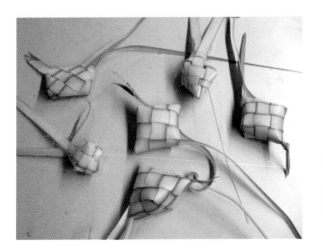

TEACHING AND LEARNING *KATUPA*

Somewhere back in distant unrecorded history, someone—or more likely many people independently—developed methods for weaving palm leaves into *katupas*. Once developed, the methods were passed on, with teaching and learning down the generations, just like all technology. And also like technology, improvements are still being made today on the core ideas and techniques that originated before anyone's memory.

I'm not going to attempt to teach you how to weave a *katupa*. It is so complex and so dependent on viewing from various angles that the process is nearly impossible to record on a two-dimensional page such as this. Although video documentation may offer some hope, the best way to learn is to grab strips of palm leaves, sit yourself right down beside the Timorese grandma, your resident expert, and attempt to follow along with her. (Don't sit across from her, for then you'd see everything as a mirror image.)

The good news is you don't have to know her language. All you have to do is become a child again and follow the leader. I always listen with interest to someone attempting to teach a *katupa*; the words coming out are clumsy and clunky: "Now stick this tip in behind that strip, no, not that one, the one coming from behind and below, and then turn it around like this and over again and back..." The teacher may as well be chanting gibberish; the words may even be distracting.

I'm convinced that learning a *katupa* and various other types of weaving happens almost entirely in the visual realm, as the learner watches Grandma's motions and mimics them. I think a lot of learning happens this way, following the more competent leader, outside any formal educational structure or institution, with no lesson plans, curriculum, or evaluation scheme. I think it stands as one of the more efficient and effective ways to learn. The lesson for a teacher is clear: never miss a chance to keep your mouth shut.

One day I was appreciating one of these hanging decorations and I realized it contained mini-*katupas* of two of the *Platonic solids*. These five solids are formed entirely from edges of a constant length and faces of a single geometric figure. The ancient Greeks found them somewhat mystic, and I do too. They are symmetric around any vertex, and it can be proven that only the five exist. Fairly regular versions of the octahedron and tetrahedron were represented in the beautiful little handicraft decoration I was viewing. Here is a close-up of three green octahedron mini-*katupas* in a hanging decoration.

I'm not sure the Timorese women who wove these Platonic *katupas* understood their profound mathematical relevance, but I've taught

many students, and they are generally filled with wonder. It's but another example of the submerged mathematics that exists in the daily life of every culture and society. Nobody calls it math, but it indisputably is.

These solids and others can be constructed through several simple methods, and then you can analyze their mystical properties. I'll show you a couple of other methods for making them at the end of this chapter, but the way we make them here is with toothpicks or kabob sticks and small cubes of old flip-flop rubber, or alternatively small cubes of potato, chayote, or young papaya. These food-based cubes are easier to prepare but often don't last as long; they turn into mold experiments unless you dry them super well.

Instead of showing you how to construct these gems, I'll lead you to construct them by following basic principles. Each of the five can be constructed by putting together uniform sticks at vertices following only two rules: the shape on each face must be always the same, and the number of sticks coming together at each vertex must also be the same. Start the table here to eventually include all the info and show more mystical aspects of these solids. Now, DON'T LOOK AHEAD! Instead prepare your sticks and cubes and start going.

Gather stuff

▸ Cubes of old flip-flop or potato, 1 to 1.5 cm on a side

▸ Toothpicks

▸ Bamboo skewers

▸ Thread

Gather tools

▸ Knife

▸ Scissors

TINKER

You can use sticks of whatever length for this, although if you use kabob sticks they become unwieldy for solids #4 and #5. To use the kabob sticks, you have to sharpen the non-pointy end in order to stab it into another cube. Toothpicks have two ends nicely sharpened.

For solid #1, stab three sticks into a cube to form a corner.

	SHAPE ON THE FACES	NUMBER OF STICKS JOINING AT EACH VERTEX	NUMBER OF FACES	NUMBER OF VERTICES	NUMBER OF EDGES	NAME
1	Triangle	3				
2	Square	3				
3	Triangle	4				
4	Pentagon	3				
5	Triangle	5				

Then put more cubes on the other ends to form triangles. Here's the first one:

Now continue linking sticks to cubes until you have a complete solid. To get the full challenge, try to do this without looking at the photo that follows. Just keep following the two rules:

- Always form triangles
- Always have three sticks coming together at a corner.

How'd you do? Did you get a little pyramid thingy like this?

Turn it around and be sure it looks the same from each side and each vertex.

There's your first Platonic solid: the tetrahedron. It's a pyramid with a triangle base. You can make many other pyramids with triangular bases that are not like this one—that is, they have longer or shorter sticks on one or more sides, making different angles at the corners, and thus don't have the super symmetry of this baby.

Now count the faces, vertices, and edges, and fill in the table:

	SHAPE ON THE FACES	STICKS AT EACH VERTEX	FACES	VERTICES	EDGES	NAME
1	Triangle	3	4	4	6	Tetrahedron

Now you're ready for the next. Again, try to cover up the following images and figure it out just from these rules! The two rules for this one are

- Always form squares.
- Always have three sticks coming together at a corner.

Start in the same way, but know that the angles will be a bit different on this, as it will turn out to be a different solid.

Now make squares on each side.

And continue making squares from three-stick vertices!

Ta-daaa! That looks familiar, eh? Technically speaking it's a hexahedron, but you can just call it a cube. Make sure it's symmetric from all faces and corners. Then fill in your table.

	SHAPE ON THE FACES	STICKS AT EACH VERTEX	FACES	VERTICES	EDGES	NAME
1	Triangle	3	4	4	6	Tetrahedron
2	Square	3	6	8	12	Hexahedron (Cube)

Now you begin to see the complexity of these things. Each of the number columns went up, but two doubled while one went by half. Who knows what lies ahead?

Continue with this process to construct the next three. On the next page are the starter hints followed by the answers. Cover the answers and try to work out the shapes just with the hints!!

3
Always form triangles.
Always have four sticks coming together at a corner.

4
Always form pentagons.
Always have three sticks coming together at a corner.

5
Always form triangles.
Always have five sticks coming together at a corner.

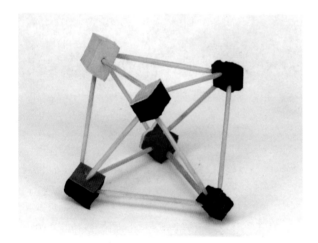

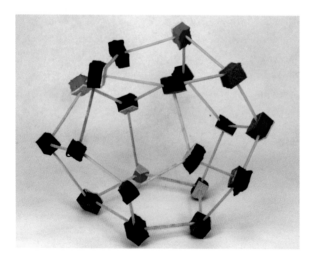

There they stand in all their glory. You know eventually you gotta have these hanging from your ceiling, but first fill in the table, then check it with ours:

	SHAPE ON THE FACES	STICKS AT EACH VERTEX	FACES	VERTICES	EDGES	NAME
1	Triangle	3	4	4	6	Tetrahedron
2	Square	3	6	8	12	Hexahedron (Cube)
3	Triangle	4	8	6	12	Octahedron
4	Pentagon	3	12	20	30	Dodecahedron
5	Triangle	5	20	12	30	Icosahedron

That's a lot to chew on. First check out those names. If they seem Greek, it's because they are. Check out the Greek prefixes for the numbers in that table:

3	tri
4	tetra
5	penta
6	hexa
8	octo
10	deca
12	dodeca
20	icosa

For geometric concerns, English uses the Greek prefixes and various other Greek language structures. *Hedron* means 3D shape, as you probably guessed.

Now compare the results in the number columns. You might have thought that as things grow more complex, the numbers in each column would increase, but in several cases, they decrease. And in some cases they *swap directly.* I put some arrows in there to emphasize that. Also note that the edge number remains the same for two pairs, which have the same background color in the table above. Cosmic.

WHAT'S GOING ON?

Polyhedra are the subject of intense academic study and have been since the time of old Plato, around 2,400 years ago. Plato referred to this set of solids in a philosophical piece about the fundamental elements of the universe, and suggested that our universe is formed from these polyhedra.

A contemporary of Plato named Theaetetus did the original proof that there are but five. You can gain some insight by trying (futilely) to create another one. Just choose a regular shape (in geometry, "regular" means all the edges are of equal length) and try to make a solid with it! Go for it! Ask your Uncle Google if you want the details.

A few words on the elegance and symmetry of the characteristics in the table: It just so happens that there is an intimate relationship between the two pairs: cube and octahedron, Dodecahedron and icosahedron. They are known as "dual polyhedra," which means that one's faces correspond to the other's vertices.

What about the lonely tetrahedron? If you morph its vertices into faces and faces into vertices you get *another tetrahedron*, the evil twin, technically called its "dual." Your good Aunt Wikipedia has got some nice animations of the metamorphosis between the dual polyhedra.

Did you notice that the ones with triangles are rock solid, sort of unmovable, whereas the cube and especially the dodecahedron felt kind of gangly, unsolid, and ready to collapse? That's the magic of the triangle, and why it's essential in construction. In the three solids with triangles, the triangles are holding the angles steady, whereas in the cube and dodecahedron, we rely on the flip-flop cube to hold the angles steady.

You don't have to stop at the Platonic solids. There also exist 13 Archimedean solids, which relax the rule about a single shape for a face. You'll learn more about Archimedes in Chapter 20. Here are two of his solids:

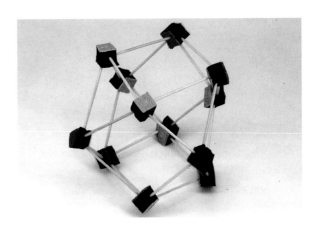

That's a cuboctahedron, made with triangles and squares. There's a different one you can make with just squares and triangle; I'll let you tinker it out yourself.

That'd be an icosidodecahedron, made of triangles and pentagons, and it's worth making if only to have an opportunity to say that word. These names are not arbitrary Greek mishmashes; they have connections to other solids that can be truncated or morphed to become these solids. If you count, you can find all 20 triangles of the icosahedron and all 12 pentagons of the dodecahedron right here in this baby. And in that previous one, all 6 squares of a cube and all 8 triangles of an octahedron. Radical, eh?

I have found the icosidodecahedron on soccer balls here, which makes sense, because it's close to a sphere. The following photo shows a bag of balls hanging in a local shop.

But look closely: those triangles are each painted on the inside of a *hexagon*! The painted design is different from actual lines on the ball! The actual design there is the Buckyball, which is found on most soccer balls. It's named after the mathematical architect Buckminster Fuller, who built domes from this and other shapes

and called them "geodesic." The Buckyball is an Archimedean solid formed from hexagons and pentagons. It's also called the truncated icosahedron, because it is formed by chopping each vertex off an icosahedron. You can make it as well from toothpicks and flip-flops!

Check out how this cheap toy soccer ball has got a pentagonal star painted into each pentagon of the Buckyball. Clever!

Then there are the Johnson solids, still demanding a uniform edge length to create regular polygons, but relaxing other symmetry restrictions. The other pyramids are examples of these solids.

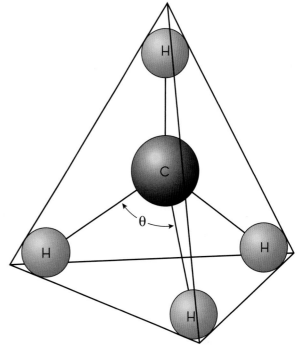

Your Aunt Wikipedia has many scrumptious photos and animations of these fabulous forms and will be more than happy to show you all the Shläfli symbols and Coxeter diagrams (haven't quite figured those out myself), as well as working out the surface area and volume of each. The website http://blazelabs.com/f-p-swave. asp (about the structure of atoms) has a great synopsis of the Platonic and Archimedean solids, all derived from the lowly tetrahedron. Mathisfun.com/geometry has interactive animations of many solids. I'm confident you'll find perusing these solids on the Internet far more enjoyable than watching another Justin Bieber video.

So that was all thrilling and soul quenching, but what's the point? Is it just abstract mathematics and a treat to the eye? Not at all, my friend. You see, various molecules and crystals have been found in exactly these shapes, including methane, the simplest organic molecule, with four hydrogen atoms positioned at the vertices of a tetrahedron:

And here is the tetrahedron *katupa*, full sized and all tied up fancy for mathematical analysis (not cooking rice). My clumsy weaving has resulted in a slightly skewed polyhedron, but with effort, this can be made perfectly regular.

Then there is the *Circogonia icosahedra,* a species of *Radiolaria,* which are single-celled organisms that make themselves complex exo-skeletons. This one is shaped like a regular icosahedron:

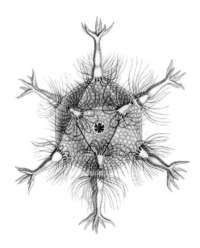

CREDIT Ernst Haeckel's 1904 *Kunstformen der Natur*

Basically, the shapes *are* sacred; they are universal forms, elegant and efficient, with symmetries so strong that nature produces them. In that respect, old Plato was right! Learning them is as enjoyable and important as learning astronomy or microbiology: this is the reality of our universe! Hang them in your room and tell all your friends to come and see!

OTHER WAYS TO MAKE POLYHEDRA

My buddies Arvind and Manish from India make geometrical solids with toothpicks and segments of rubber bicycle valve tubes available widely in India and China (but not the United States!). You have to make holes in the tubes at the joints and thread through another tube.

I've made them with straw sections and pipe cleaners—excuse me, "chenille stems." You just wrap together the wires at the joints. No reason you shouldn't have a few dozen of these hanging around the house as well.

Part III
Life and Living

Chapter 10

FOOD CALENDARS

Chart what you eat and where it comes from.
CREDIT: Thanks to Peg Vamosy for this photo of Mana Joana and her students harvesting rice in the Aileu district.

In the United States every school child knows that milk comes from the supermarket, and in fact food in general comes from a box, bag, or can. Fresh fruits and vegetables are also available; heck, most of them are available consistently, month after month, year-round. They've got tiny little tags that tell you grudgingly where they're from, but who cares? Pass the celery! That all this life-giving food was produced on a specific patch of earth somewhere is a fact lost on a great many children, and even adults.

In Timor most people know where the majority of their food is produced. Not only do subsistence farmers know that it comes from the earth, they also know which patch of earth produced each calorie, in what month, and with what

additional inputs of labor and material. Just as young American kids learn their shapes and colors as toddlers, farmer kids in Timor learn about their food.

This is no mystery. Food is one of the key elements of life, like air and water, and when it is hard to find, life becomes no fun at all. In the United States, industrial food systems have made it seem like food is a sure thing, a permanent fixture on the supermarket shelves, polished and perfect, packaged for long life and easy transport. Even if you run out of food in your house, you know it's not because there is none in the store down the street.

Doing this activity will be a challenge if you're not living in the Majority World, but it should give you a good new perspective on how you feed yourself. The reason we teach the food calendar here is because there are still sections of the year in most of Timor-Leste when not much is coming ripe from field or forest. Some things are available year-round, such as bananas and papayas (these can be used as vegetables as well as fruits), and meat and fish have no particular season. Still, there are times of the year in many locations referred to as the "hungry season." Government and local organizations are working to help farmers look at all options to fill in the sparse parts of the calendar with crops that will create a welcome harvest during those times.

If you don't have the problem of a hungry season, then you chose your parents well, and the industrial food system is keeping you satisfied, at least temporarily. But take this chance to consider many in the Majority World who understand, as we all really should, how much we depend on the very soil beneath our feet.

TOWARD THE GOAL OF FOOD SOVEREIGNTY

Industrial agriculture has made incursions into Timor in the form of seed varieties and pesticides as well as imported foods in the local markets, such as Chinese garlic, onions, and potatoes; Indonesian instant noodles; boxed milk from New Zealand; and juices from the Mediterranean. Each of these has benefits and liabilities. It's nice having potatoes year-round, but as farmers and consumers become reliant on imported products, they lose their independence and slowly forget the methods that have worked here for hundreds of generations.

Many farmers are leaning hard toward sustainable local agriculture. The Sustainable Agriculture Network for Timor-Leste (HASA-TIL) promotes and supports sustainable and organic methods that prioritize food sovereignty, putting farmers and communities in charge of their own caloric intake, now and for generations to come.

In the newly developed national primary school curriculum, the Ministry of Education included a lot of information about permaculture as developed by Permatil, Timor-Leste's local permaculture organization. Check out their Facebook page: Permakultura Timor-Leste.

The goal is for schools to all have organic gardens, which students and teachers will use to learn sustainable agriculture as well as botany, entomology, mathematics, and more.

CREDIT: PHOTOS FROM THE SCHOOL GARDEN IN OBRATO MANATUTU THANKS TO EGO LEMOS!

TINKER

Write down the 10 meals or snacks you eat most, day by day or week by week. Don't write down individual foods, but rather the main foods of a meal, or the main group of food for common snacks. We'll keep this simple by ignoring drinks. Here's a hypothetical list that could be the diet of my niece, living near Kansas City, Missouri:

Hamburgers with fries

Spaghetti

Sandwiches

Cereal

Burritos

Chips, crackers, and cookies

Apples, oranges, and bananas

Pizza

Pancakes

Donuts and pastries

Now see if you can dissect those foods into their elemental ingredients. Don't worry about the small stuff like sugar, salt, spices, and oil, but try to get all the major foodstuffs. For example:

Hamburgers

    Wheat (bun)

    Beef

    Cucumbers (pickles)

Potatoes (fries)

Spaghetti

    Wheat (pasta)

    Tomatoes, canned (sauce)

    Beef (sauce)

    Garlic (sauce)

    Mushrooms (sauce)

Sandwiches

    Wheat (bread)

    Peanuts (peanut butter)

    Strawberries (jam)

Cereal

    Corn (grain, not sweet corn)

    Milk

Burritos

    Wheat (tortilla)

    Beans

    Chicken

    Rice

    Lettuce

    Tomato, fresh

    Onions

Chips, crackers, and cookies

Potatoes

Corn (grain, not sweet corn)

Wheat

Apples, oranges, and bananas: no mystery here

Pizza

    Wheat (crust)

    Tomatoes, canned (sauce)

    Beef (pepperoni)

Pancakes

    Wheat

    Eggs

    Milk

Donuts and pastries: Wheat

Now regroup the elemental ingredients into a big table, each one listed only once, and separated roughly into categories. On the right, leave space for the 12 months of a year:

		J	F	M	A	M	J	J	A	S	O	N	D
Meat, dairy, and eggs	Chicken												
	Beef												
	Milk												
Grain and other starch	Corn (grain, not sweet corn)												
	Wheat												
	Rice												
	Potatoes (frozen)												
Veggies and legumes	Lettuce												
	Tomato, canned												
	Tomato, fresh												
	Onions												
	Cucumbers												
	Mushrooms												
	Beans												
	Garlic												
Fruit	Apples												
	Oranges												
	Bananas												
	Strawberries												

Tomatoes appear twice because they were fresh on the burger and canned and processed in the spaghetti sauce. This differentiation will help later on.

Now that we have most of our main foods in the calendar, we'll mark it up get a better perspective on what we're eating. Here is the gist of the system we'll use:

Green = fresh food that may have been grown locally

Blue = food that was preserved (frozen, canned, processed)

Orange = fresh food that was definitely shipped in from far away

X = harvest season for a given food, meaning it could have come straight to your table from the farm

Follow me closely, doing it for your home, whereas I'll be doing it as if for my niece in the Kansas City area. You can check the marking for each step on the final calendar later in the chapter.

Put an X on months for harvest seasons on those items that could have been produced within a 100-mile (150-km) radius of your home. This is sort of arbitrary, but it gives us a baseline to work with. This is a tricky part for those of us no longer connected to the land. You may have to ask someone you know with a garden, look up harvest calendars in the library or online, or make your best guess. Put an X on the months when each of these foods would come ripe or ready, *if they were produced where you are.*

Before we go on, we have to accept that just because these foods were ripe locally doesn't mean that the ones you ate were actually the local ones. That's the way our convoluted, industrially opaque food system works now. In California, we live just up the road from Castroville, the artichoke capital of the nation, yet routinely see artichokes from Spain for sale in local stores. My friends were biking through the colossal wheat fields of Kansas, yet in the small-town mom-and-pop shops they stopped in for supplies there was nothing but corporate white bread shipped in from hundreds of miles away. That happens all the time. It's a good reason to find out more about what you're eating and where it came from.

Anyhow, these Xs represent a possibility: that this food could be found fresh in that month where you are.

We're already starting to notice a distinct pattern in the calendar. It's clear that animals can be milked or slaughtered any time of the year, but most crops only grow well in the summer throughout most of the temperate region, and take weeks or months to get to harvest readiness.

Now, for foods that are not grown locally where you are, mark the whole row orange, but not the food's name.

After that, highlight food names in green for each food that you ate fresh or stored/preserved naturally (dried, kept in cellar) and not canned, frozen, or otherwise processed. By *processed* I mean it was altered in some significant way. For example, wheat was made into flour, but all the

original stuff is there (in the case of whole wheat). Rice was hulled and maybe polished, but you're still getting the real stuff. These both count as fresh—mark them green. The potatoes that made your fries were frozen and the milk was homogenized and jugged, so that counts as not fresh, as does the frozen meat. If you had freshly slaughtered meat and chopped up fresh potatoes for your fries, put 'em in green.

How much green have you got? That's a big distinction right there. Here in Timor, many people's calendars will be nearly 100 percent green. Exceptions will be canned fish and other meat, and instant noodles. Most Timorese simply do not think of the supermarket when they think of food. They think of their fields and gardens.

Before the next step, we need to talk about storing and preserving food. Just by looking at the Xs on the calendar, you can see the problem. It's the same problem faced by our great-grandparents 150 years ago all the way back to our ancestors 6,000 years ago when agriculture was just starting up: What's to go with the meat for supper when those other crops are not ripe yet? In other words, how to preserve these crops so that we can eat them for months after they've been harvested? Canning and freezing are modern ways, and a range of ancient ways, such as drying, salting, pickling, and cellar storage, are still used today.

Here we'll make a distinction between the more modern, processed methods of freezing, canning and packaging (with blue), and the old ways (with green).

Highlight names and entire rows in blue for those that you ate preserved (frozen, canned, or processed). If the row is already orange, meaning it came from afar, leave it. This blue distinction is made according to how the food is manipulated before it gets to you, but you're also often hard-pressed to know where these were produced. Some may have been local, but others came from outside the country, maybe the other hemisphere.

Highlight all squares with Xs that are not blue or green, and also highlight in green the squares for the months that these items would last if stored/preserved naturally.

You can see a few foods are made to last: grains dried will last the year, and many root crops as well as some fruit will keep for months in cool areas. Other wimpy veggies like lettuce last only days once harvested, even with refrigeration.

Highlight all remaining squares in orange, meaning these were definitely shipped from afar if you ate this stuff during this month.

There's your food calendar!

		J	F	M	A	M	J	J	A	S	O	N	D
Meat, dairy, and eggs	Chicken	X	X	X	X	X	X	X	X	X	X	X	X
	Beef	X	X	X	X	X	X	X	X	X	X	X	X
	Milk	X	X	X	X	X	X	X	X	X	X	X	X
Grain and other starch	Corn (grain, not sweet corn)								X	X	X		
	Wheat				X	X	X						
	Rice												
	Potatoes (frozen)								X	X	X		
Veggies and legumes	Lettuce						X	X	X	X			
	Tomato, canned						X	X	X				
	Tomato, fresh						X	X	X	X			
	Onions								X	X	X		
	Cucumbers							X	X	X			
	Mushrooms												
	Beans							X	X	X			
	Garlic									X	X		
Fruit	Apples							X	X	X			
	Oranges												
	Bananas												
	Strawberries				X	X	X	X					

Here's the list and calendar for our family, mostly vegetarian, here in Timor-Leste.

Rice, beans, tempeh, tofu, and vegetables

Pasta with sauce and cheese

Fish and chips (fries)

Pizza (homemade)

Oatmeal and once-a-week cold cereal

Toast with butter and jam or honey

Cheese sandwiches

Banana chips, peanuts, crackers

Fruit, mostly local

Salads

Here's our calendar:

Category	Food												
Meat, dairy, and eggs	Fish	X	X	X	X	X	X	X	X	X	X	X	X
	Milk												
	Eggs	X	X	X	X	X	X	X	X	X	X	X	X
Grain and other starch	Rice	X	X	X	X	X	X	X	X	X	X	X	X
	Oatmeal												
	Wheat												
Veggies and legumes	Kanko (water spinach)	X	X	X	X	X	X	X	X	X	X	X	X
	Carrots	X	X	X	X	X	X	X	X	X	X	X	X
	Lettuce	X	X	X	X	X	X	X	X	X	X	X	X
	Mustard greens	X	X	X	X	X	X	X	X	X	X	X	X
	Eggplant	X	X	X	X	X	X	X	X	X	X	X	X
	Tomatoes, fresh	X	X	X	X	X	X	X	X	X	X	X	X
	Tomatoes, canned	X	X	X	X	X	X	X	X	X	X	X	X
	Cucumbers	X	X	X	X	X	X						
	Garlic		X	X	X	X	X	X					
	Onions		X	X	X	X	X	X					
	Pinto beans		X	X	X	X	X						
	Peanuts (PB)		X	X	X	X	X						
	Potatoes				X	X	X	X	X				
Fruit	Avocados	X	X	X	X	X						X	X
	Banana	X	X	X	X	X	X	X	X	X	X	X	X
	Passion fruit	X	X	X	X	X	X	X	X	X	X	X	X
	Papaya	X	X	X	X	X	X	X	X	X	X	X	X
	Strawberries (jam)								X	X	X	X	

Check out all those Xs! All that green! Welcome to the tropics! I've labeled green anything that's produced on this small half-island, calling it all local. A couple of small operations are producing jam from the high-mountain strawberry operations, and I'm here to tell you, they know what they're doing.

You can see that if we could break our wheat, oat, and milk habits, we'd be pretty much eating all local. Of course we eat other stuff: jams and fruit juices and junk food treats mostly from Indonesia, but our main sustenance is represented in the calendar.

Here is our calendar cut out and formed into a hanging spinner mounted on a bent hanger nailed to the wall. It drifts around, a bit like the earth and the seasons, to remind us about the sources of our food.

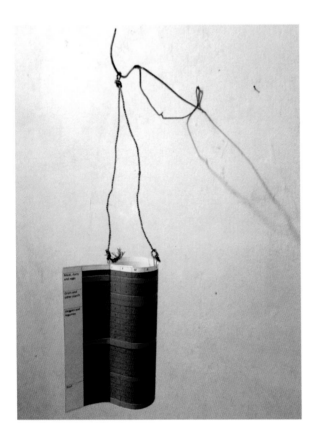

WHAT'S GOING ON?

Let's go over the meaning of these markers again:

Green: Fresh, and produced locally so you may be eating fresh local (though not at all a sure thing)

Green X: Could be straight from a nearby field

Blue: Preserved, and who knows where it's from

Blue X: Missed opportunity—those months you could have eaten this item fresh local, but chose the preserved, probably shipped in

Orange: Either shipped in from afar or produced in a greenhouse for sure

Greenhouses, by the way, can function as a local, ecological option to lengthen the season of fresh produce. The greenhouse effect can keep the temperature up so that various vegetables can grow inside when they would die outside. On the downside, some places also burn petroleum to heat greenhouses and produce things like tomatoes when it's snowing outside. So you may hesitate before indulging in those hothouse tomatoes when you're wearing four layers. If they seem a bit expensive, it's probably due to the massive heat loss through the thin walls and roofs of the greenhouse.

The main modern method of accessing "fresh" out-of-season foods is shipping them in from a place where they are in season. A lot of crops like lettuce and tomatoes can be grown across the southern strip of the United States pretty much any time of year with enough irrigation, and the pear season in Chile just happens to be the dead of the Minnesota winter. If the shipping can be carried out quickly enough, we still get fresh food in the supermarket. Of course, agro-industry will do anything to keep that food looking fresh longer, including dumping chemicals on it and breeding it in ways that deprioritize taste, so let the buyer beware. And of course all that transport has serious ramifications for climate change. So it's good to stay a bit skeptical when you see fabulous stuff in your supermarket out of season in the fresh section. Local fresh food is really the only fresh fresh food.

One weakness of our calendar is that it assumes you eat more or less the same meals year-round. You may not, and your ancestors surely didn't, and Timorese don't either. Folks here were and are mostly dependent on the growing and gathering seasons of their climate.

The key to sustainability is eating local stuff. You want to maximize green in your calendar and make sure blue items are produced locally. The extent to which you can do this will depend on your flexibility and the diversity of agriculture where you are. You will need to eat more with the seasons, like your dear great-great-grandparents did. Ironically, if you are keen to eat more local foods, one of the key solutions is to rely on foods preserved with both older and modern methods. You could do canning yourself, or buy a large freezer for garden excess. But you can also find and patronize local businesses that preserve local food.

Where is the most orange? In the Kansas City calendar shown here, it's in April and May. Those were the hungry months of our temperate climate ancestors: birds singing, meadows in bloom, crops bursting forth from the soil, but *not...yet...ripe*. So temperate climate peoples through the ages have had to wait and continue

eating the half-moldy stuff in the cellar from last year's harvest.

You can imagine the extraordinary amplification of appreciation when the first vegetables came ripe after a period of waiting—the first tender peas to fill out their pods, the first crisp radishes to redden, the first berries from the vine. Mmmm, it makes my mouth water just fantasizing about that moment.

Aside from being pleasurable, eating with the seasons spares the world the waste of transport emissions and supports local farmers. It may be interesting for you to try eating nothing but local for a time—one week, one month. Barbara Kingsolver tried it for a year, as you can read in her book *Animal, Vegetable, Miracle*. Permaculture's magazine and website are inspiring. And just strolling into some dirt and planting something is like a trip to the ancient past; the essence of our existence as an agricultural society comes down to seeds growing in soil to produce something that will sustain us. Folks here in Timor don't have any doubt about that.

ACCESSING FOOD AT DIFFERENT LATITUDES

In high-latitude locations, sometimes preserving food is easy. When I was living in China, I took a trip to Inner Mongolia for Chinese New Year in January. The train cars were heated with coal and everyone got two thick quilts in my sleeper section, but I was still stiff with cold. Imagine my shock when the vendors came down the aisle selling not hot chocolate, not hot chestnuts, not hot rolls, but *popsicles*, sitting on their cart, no refrigeration necessary. I declined.

Upon our arrival in Hailear, we walked through the streets and saw vendors selling frozen fruits and vegetables at their stands—again, no cooling necessary. The food had simply frozen on the trip north. Rock-hard frozen pears from southern China were popular; they were almost like a fruit pop. Then I noticed nearly every apartment had bundles hung out the windows: free freezers for the masses. Indeed, it never got anywhere near freezing for the week I was there. The astonishing thing I realized was that all of Siberia was still northward from us, and still colder.

In the tropics it's harder to preserve food with simple methods. Things rot fast; cooked rice or beans left out overnight go bad by the next morning. Corn is often hung over the fire so the smoke keeps the bugs and mold off, as in the photo here. A local aid group here donates sheet metal bins to farmers to help them do battle with the bugs and rodents, but the grain has to be very dry before it goes into the bins or it will rot inside them.

I've also noticed that pickling is not as prevalent here. Subsistence farmers I knew in China would pickle a portion of each harvest, often in traditional pottery that did not require modern seals. Here there is no such parallel, and I think it makes sense. It may be harder to get the pickling process to counter the rampant decomposition process here in the tropics, but there is

also less need for it. When the ice is blowing in northern China and no food is to be found in the snow-covered garden, the next season's crops are just coming ripe in Timor.

CREDIT: Luis Nivio Soares

It's a happy happenstance of nature that at just those tropical places where it's hard to preserve food, there can be food coming ripe for harvest nearly year-round. In Timor it's common to hear folks talking about the "corn season," the "Time of New Corn" (meaning the first of the new harvest, as opposed to the old stuff they'd been eating for months), the time of planting rice or harvesting rice, and so on. Mestre Caetano says his grandfather used to answer the question, "When were you born?" with "In the corn season." Life literally revolves around the harvests.

Chapter 11

TINKERING WITH PLANTS

Get plants to grow in funny ways.

ear Watsonville, California, there is an amusement park with some trees that were tinkered by a very patient man into astonishing shapes: circles, squares, grids, telephone booths, spirals, arches, and more. Ask your Uncle Google about Gilroy Gardens Circus Trees. While I was inspired when I saw these, I knew I didn't have the patience to see one of my own through to such a marvelous result. Thus it was that I began tinkering with faster-growing stuff. This chapter displays some of our recent efforts here in the tropics.

Gather stuff

- Seeds, cuttings, shoots, whatever you think may grow
- Pots, cups, plates, boxes, whatever you think the plants may grow in
- Scrap wood, metal, plastic, whatever you need to coax the plants to grow the way you want
- Dirt
- Water

TINKER

My teacher buddies here use mung bean sprouts to do germination and growth experiments because they really shoot up in the first few days of their lives. So we also try to get some to weave themselves in interesting ways around a lattice. We plant them in a plate of dirt under the lattice. Here is the most recent run:

I recently reached for the string bean seeds. Here's what I got:

See them going in and out of the holes? That all takes place within a week and a half and is pretty cool, but then they often stop growing. You can also do this with PVC bits. We wait until the sprouts emerge and then fasten each tube down over a sprout so it has little option but to venture up the tube. In the photo here, all but one came out, and it had the hardest road: a 90-degree angle after a long straightaway. Its first leaf is 1 cm inside the opening.

I find these mung bean sprouts speedy but fragile. Looking for something more robust,

You can see them looping around the horizontal beams quite nicely. It's been about three weeks now and that's where they are as of this writing. I expect to be looping them around for a few more weeks until they get tired and start producing beans.

I noticed a vine growing up the side of our house. I got a tube and stuffed it partway down. It knew what to do.

and that shaped its future. We can do the same thing with a papaya, only with a glass jar. You have to break the jar to get it out. This one is filling out quite nicely in its square prison.

You can also alter the future of shoots and trees. This is our new banana sprout and another baby tree in our front yard. I just gave them rings: a piece of PVC and a bottle cap with the top cut out. Fashion accessories, you could call them.

This opens up all sorts of possibilities. I bet I can make a veritable spider web with this maverick vine, given time.

I found photos of square pumpkins and watermelons on the Internet. The creative farmer had built a wooden box around a baby pumpkin

It's fascinating to watch the plant deal with these bangles. Both of the rings in these photos have been in place for more than a month, and the trees seem to be accepting them well. The tree with the PVC will not press on it for years probably, but the banana tree has already begun stretching the ring and squeezing itself.

(That big ugly stalk next to the banana spout with the bottle cap ring is its mummy. It's around 6 meters tall, two years old, and we're still waiting for the bananas to come. Bananas usually don't reproduce with seeds, but rather send out shoots like this that result in more plants right around themselves. Redwoods can do something like this too!)

Another thing we've done that's easy and instructive is to tie knots in plants. Anything that's limber enough can be twisted on itself to become a simple knot, and then monitored to see if it keeps growing and in what manner. This chapter's introductory photo is a coconut frond all dolled up with decorative knots, which it seems to be reasonably happy with. Two more we're currently monitoring are here:

I'm sure you can think of other creative ways to get plants to do crazy things. By the way, this is clearly a lazy person's tinkering technique: 5 minutes of action, many months of pleasant observations.

WHAT'S GOING ON?

The issue of plants growing is fascinating in and of itself. We animals have muscles that pull on bones and move us around. Plants don't have muscles. Plants have cells with stiff cell walls—that's it. Any growth or movement that they do is by multiplying or changing of shape of these cells. A number of plants in Timor make movements. One, called "Shy Maria," closes its little array of parallel leaves if anything touches it. Another, which looks a lot like the one I ringed earlier, is called the "Sleepy-eyed tree," and closes up its leaves tightly every night.

How do they do this? It's the same question as for carnivorous plants: How does a Venus fly trap close so quickly and catch insects without muscles? The details are complex and I'm no expert, but the answer is the same: by multiplying or shape-shifting cells.

I'm impressed with the speed of many plants here. That papaya tree in an earlier photo is 5 meters tall and scarcely over one year old. Look at those fat papayas hanging up there. Papaya trees grow fast because the trunk is full of air. The wood is more like a honeycomb than a solid piece of pine or oak. When it dies, it decomposes into compost within a few weeks.

But other plants grow for a long, slow time. Teak wood and sandalwood trees take decades to mature. The speed of each plant is related to its priorities for life, which are dictated by its environment and have evolved and adapted over millions of years. Basically, all life around you is a brilliant success story. The failures are only visible as fossils. So enjoy your plants as you tinker, and remember: you need them!

Chapter 12

FINGER MODEL AND BIRD FOOT DISSECTION

Tinker the hand.

Have you ever really looked at your own hands? They are so elegantly complex that it blows the mind. Check out all the levers and cables under your skin and the fantastic number of different movements you can get out of that system. Here we'll make a super-crude model of the system to enact just the simplest of motions.

Bird feet are similar in complexity, and we can take them apart, since someone else is going to eat the rest of the body. Over here, it's not so common to eat the feet, but there are plenty of Chinese here, and they love chicken and duck feet, so it was not hard to find these at the store. You can do this tinkering with any bird feet that you find.

Gather stuff

▸ Bird feet

▸ Plates or cardboard to work on

Gather tools

▸ Razor blades and/or sharp knives

▸ Magnifying glass, if you have one

TINKER

Start by hacking away the skin up near the top of the leg stump. You're looking for white tendons, a whole wad of them.

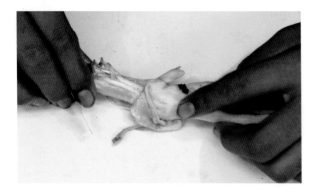

When you can isolate one, pull on it while holding the main part of the foot. See what moves.

Each one of those white tendons is connected to one of the tiny bones making up the toes of the foot. If you pull them all, the whole foot should curl up.

Sometimes they seem fused together, but try to get them apart for maximum movement. When you pull on one, you'll see which toe comes up. Then you can start stripping around that toe to follow the tendon down. Get out your hand lens if you have one and inspect things close up.

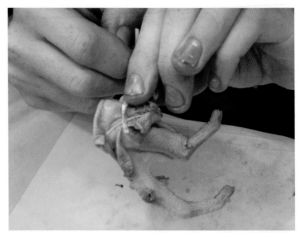

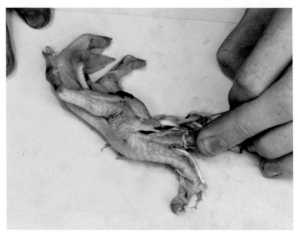

You should be able to get all sorts of movement by pulling on different tendons.

The ones on the top of the foot will make the foot lift up.

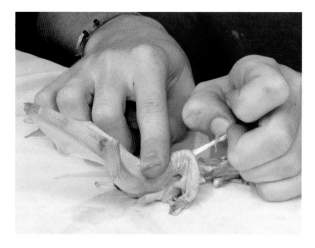

Keep on stripping off skin and separating tendons until you can get each part of each toe to move.

WHAT'S GOING ON?

What you're doing is just what the bird does when it walks, with your hand taking the place of the bird's muscles. Bird feet are a lot like ours in that sense—not too many muscles down inside there. Muscles are up in our calf or the bird's leg, and the force gets transferred to the bones by means of these tendons.

In biology, virtually every structure is a result of competition and adaptation, and ultimately evolution. The segmented structure of our hands and feet is similar to most other mammals as well as plenty of birds, reptiles, and amphibians. It's clearly a structure that works in a diverse set of circumstances. From grasping branches to clawing through the duckweed to pulling tendons in a duck's foot, segmented digits get the job done.

Marine mammals are an example of where it failed. Check out the arms and hands of a whale, dolphin, or seal, and you'll see they've got most

of the bones we have: single upper arm bone (humerus), double forearm bones (radius and ulna), and the three types of hand bone (carpals, metacarpals, and phalanges). But they barely use them! They're all stuck together into a big fin that just flips back and forth in the water. That's essentially why you've never seen a walrus playing piano—just wouldn't work.

This was a case where the structure was not useful for these mammals that evolved back into the water, so it just sort of got put aside by evolution. Two little bone nubbins on each side halfway down are all that are left of the whale's leg system; they're called the vestigial pelvis.

But getting back to the tendons of hands and feet, robots arms can be made that use a similar system. Let's make a crude model to see it in action.

Gather more stuff

▶ Straws, not too thin, no bendy part necessary

▶ Zip ties

▶ Small string like kite string or thin wire

▶ Narrow transparent cellophane tape

▶ Craft sticks or tongue depressors

Gather more tools

▶ Scissors

▶ Side cutters or hefty scissors to cut the sticks

TINKER

First we'll make a super-simple version, which I got from my buddy Sam, on Instructables. He's the mind behind The Oakland Toy Lab, with around 100 fine Instructables, but undoubtedly he nabbed this project from someone else, good tinkerer that he is. Bend the straw wherever you want a joint and cut out a little triangle.

Cut two or three of these along your straw, leaving one end without any cuts. You're looking to cut about halfway through it so that when you open it up again, it looks like this.

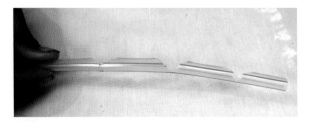

Now slide a zip tie down through the straw so that the tip comes out the end with the holes. Bend the zip tie tip over the end of the straw and tape it on securely.

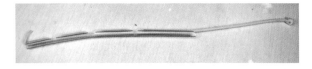

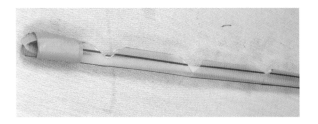

You're done. Hold on to the non-taped end and yank on the zip tie. Way cool finger action.

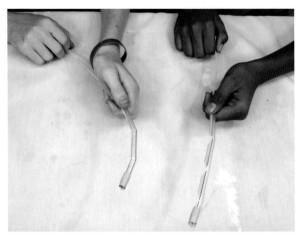

So that was pretty easy and fun, eh? It just happens to be anatomically incorrect. Sorry about that. Hey, I wanted you to have the fun before I broke it to you. Look: bend your fingers into a fist and then let your hand go limp. Do your fingers spring back up like that happy straw? Methinks not. If you want them to stretch out again you have to pull them back up. So let's make another model that's closer to reality.

Start with the bones: craft sticks. You can use larger craft sticks or tongue depressors if you want to. Cut them into pieces about 5 cm long.

The point is to have some segments that are solid like bones. Leave one long to hold on to.

Now cut some straw segments. You're going to put these segments on both sides of the sticks. The straw segments on one side have to be exactly the same length as the sticks, and the ones on the other side have to be slightly shorter.

Before you go any further, tape these segments together. Run a strip of cellophane tape right down from top to bottom, then turn it over and do it on the back too. Have the sticks touching, but just barely. When you're done, the segments should be held pretty tightly but still be able to bend.

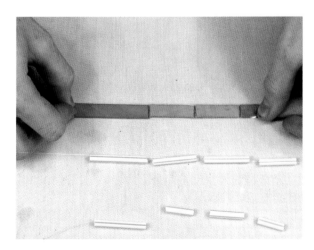

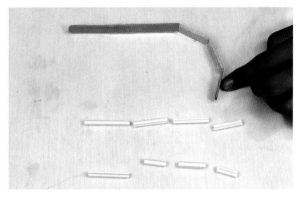

Fasten the straw segments onto the sticks now. You can just tape them, but an easier way is to use hot glue. Careful! We use this high-level tinkering technique: make a little pool of hot glue, and then quickly dip the segments in one by one and stick them to their respective sticks, front and back. The longer ones should be touching end to end, the shorter ones centered on their sticks.

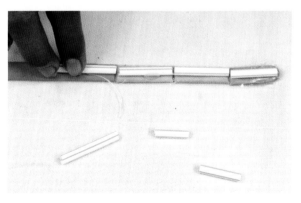

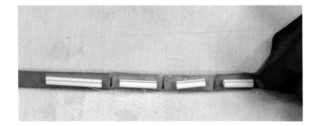

When they're all sitting well in place, reinforce them with tape.

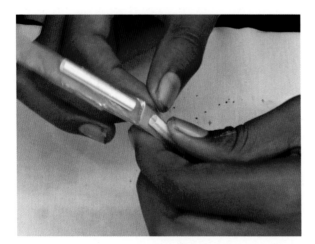

Now thread thin wire or string up through the straws on both sides. Tie off the end to the tip or to the tip straw. Leave the other end of the strings a bit long to hold on to and make little knots or stirrups on that end.

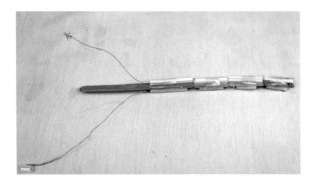

Now you're ready for some real finger action. Hold it by the long stick and gently pull each of the wires.

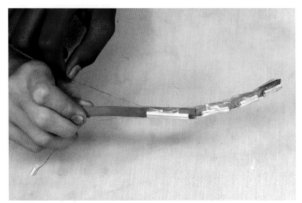

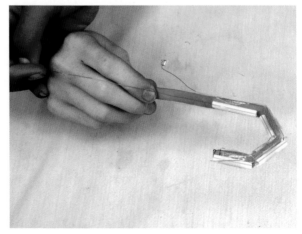

You see that the longer straw segments on the back line up and run into each other when you pull on their string, while the short ones on the front allow the finger to bend because they have gaps between them. You can tweak your finger by shaving off straws or adjusting their placement.

Now this one is like your finger, right? You let go and it just stays where it is!

WHAT'S GOING ON?

In this model the sticks are the bones, the wire is the tendons, your pull is the muscle, the tape holding the sticks together are the ligaments, and the straws are the tendon sheaths.

A famous tendon sheath is the carpal tunnel. Heard that word before? Some people do the same action over and over like typing or playing the banjo, and the wrist tendons and their sheaths in the midst of that motion begin to get inflamed and sore. It's called Carpal Tunnel Syndrome, and I suffered through it when I was a child. (I didn't actually have it; my mom had it, but when she got her operation, she couldn't get her wrists wet so my brother and I had to do all the dishes for nearly two months. It was grim.)

Muscles in your forearms pull on the string-like tendons that then pull on the bones. It's a great system until it goes wrong. (My brother mowed into the back of my heel with a lawn-mower when I was in third grade and sliced halfway through my Achilles tendon, the big one above the heel that puts a spring in my step. I've forgiven him.) Tendons get injured and heal fairly routinely. Both tennis elbow and shin splints are tendons acting up. If you take care of them and exercise, they'll cure themselves. Ligaments don't heal nearly as easily; if you tear a ligament, you've got serious problems.

Another thing this model shows is the truth about the way muscles impart force: they always pull, never push. Even when you're pushing, it's happening by means of muscles pulling on the levers of your arm and hand bones. As you pull your two wires, remember that you almost always find two: one to make the motion and one to retract.

See if you can think of other parts of the musculoskeletal system to make models of—we've done arms, legs, and vertebrae as well!

Chapter 13

HEART MODEL AND HEART DISSECTION

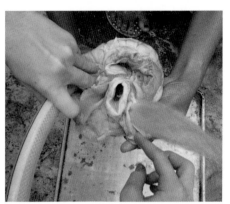

Tinker the heart.

Your heart is one of the more important muscles in your body. The heart starts beating when we are an embryo in our mother's womb, only three weeks after the moment of fertilization between the sperm and egg. When it stops, you stop. In fact, that's a key part of the definition of death: the old ticker is not pumping anymore.

The heart's got an astonishingly intricate structure, and you can see it all clearly if you dissect the heart of our fellow mammal, the cow. You can check out other hearts, too, but the cow heart is nice because it's so much bigger than ours, which makes sense because it's been pumping blood around a much bigger creature. First you should make the model, though, as it's simpler than the real thing. When you understand the model, I'll show you the dissection.

Gather stuff

- Dish detergent bottle with an oval cross section (in other words, not round)
- Thin, clear tube, around 6 or 8 mm (¼″) diameter, 50 cm long
- Coin that fits in the neck of the detergent bottle and blocks the liquid going out
- Balloon, on the big side
- Rubber bands
- Red coloring for the water
- Basin
- Water

Gather tools

- Scissors
- Drill and bits, or nail

TINKER

First, get the label off your bottle so that you can see the movement of liquid inside it. You don't have to clean out the bottle; a bit of soap makes bubbles, which help to show the motion of the water.

Put a hole in the side of the bottle, about one-fourth of the way up from the base. Make the hole just smaller than the thin tube you'll be putting in here. This is a somewhat critical step to this project. If you make it too big, it will leak air or water, and once it's wet, you'll never get glue or tape to stick to it. So take your time and get it right the first time. We started with a small drill bit and then reamed it out with scissors held perpendicular. You can also just start with a nail hole and then move to the scissors.

Enlarge the hole tiny bit by tiny bit, always keeping it a perfect circle, until you can just barely poke one end of that tube in, *without pinching it off*. It should be hard to get in, but jam it in maybe 8 cm, curving downward within the bottle upside-down, like the photo here:

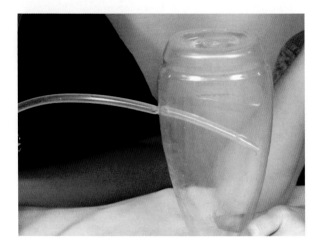

Now cut a tiny hole in the side of the balloon. You just barely have to nick it.

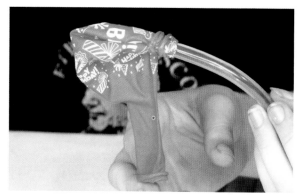

Poke the free end of the clear tube into the balloon through this hole, and then use a rubber band to secure the balloon onto the tube. Here is another touchy part: the rubber band should be tight, but not so tight that it squashes the tube.

Again, do your best to get that joint right while it's still dry. Once it's wet and soapy, it'll be hard to work with.

Drop the coin into the opening of the lid. It should sit down flat over the holes of the spout. In the photo is a Timorese 5 cent piece, but I bet pennies would work just as well.

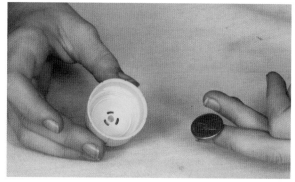

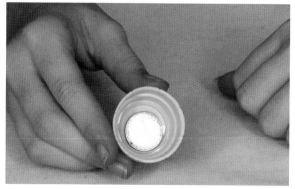

That's just to check its size. When you're confident it's right, chuck the coin into the bottle and put the mouth of the balloon over the nozzle of the bottle. You can secure this with another rubber band if you like, but sometimes it's tight enough without a rubber band.

You're ready for action.

Fill the bottle up about three-fourths; when turned over, the water level should be just about to the hole on the side. Then put some red color in to make it look like blood. If you don't see much foam, you can add a bit more soap.

Now put the lid on firmly, straighten out the balloon, tip it all over, and start trying to pump some blood.

2. Squeeze the tube near the place it enters the bottle on the side.

Here's the sequence for pumping. It's a four-step cycle:

1. Squeeze the bottle.

Water stops. Balloon is full.

3. Let go of the soap bottle. Keep squeezing the tube and try not to let any water go back into the top of the bottle.

Water comes out of the tube near the top and flows down the tube and in through the side of the balloon.

Water flows into the bottle, up through the spout at the bottom and through the hole with the coin. You should see some bubbles coming up.

4. Let go of the tube.

Water stops. Balloon is limp.

Then do it again, and again, and again! You're pumping blood, baby!

WHAT'S GOING ON?

Humans and other vertebrates have a circulation system that handles transportation in the body. It's composed of organs like the heart and blood vessels (arteries, veins, and capillaries) and of course plenty of blood. Food substances, hormones, and oxygen get delivered to the whole body, every single cell, and on the same trip carbon dioxide and the wastes from cell processes get taken away.

The heart has the key role of pumping blood on this transport circuit. Humans and mammals (cattle, pigs, and goats, for example) all have the same sort of circulatory system and the same heart structure. It's got four cavities: right and left atrium and right and left ventricle. Wait a minute, you may say, the model we just made only

has two cavities. That's right; technically you just made a model of one side of a heart. Check out this diagram if you want to get all the names of the little parts.

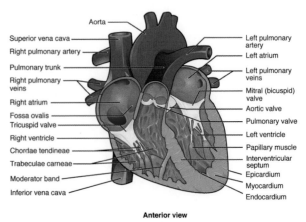

Anterior view

CREDIT: OpenStax College

The ventricle is the pumper, whereas the atrium is sort of an elastic holding space for the blood when it's coming back around. Between the atrium and ventricle on both sides are valves. These are one-way valves. They prevent blood from going backward into the atrium when the ventricle pumps. There are also two other valves at the base of the arteries, with the function of preventing blood from going back into the ventricles once it's been pumped out.

The heart works as follows:

1. The two atria enlarge.

   ▶ Blood from the body, rich with carbon dioxide, flows into the right atrium.

   ▶ Blood from the lungs, rich with oxygen, flows into the left atrium.

2. The two ventricles enlarge and pull in blood from the two atria.

   ▶ Blood from the right atrium enters the right ventricle.

▸ Blood from the left atrium enters the left ventricle.

3. The two ventricles contract.

▸ Mitral and tricuspid valves close, so blood cannot return to the atria.

▸ Blood from the left ventricle, rich with oxygen, is pumped to the whole body.

▸ Blood from the right ventricle, rich with carbon dioxide, is pumped to the lungs to be renewed.

4. Semilunar valves close so that blood cannot return into the ventricles.

5. The process begins again and repeats itself.

Blood flows to the entire body through tubes called *vessels*. Three kinds of vessels carry out the different roles:

▸ *Arteries* are the tubes with thick walls that transport blood from the heart and take it to every tissue in the body.

▸ *Veins* are tubes with thin walls that transport blood back to the heart.

▸ *Capillaries* are the tiny tubes that link the arteries and veins and are the places where water, oxygen, carbon dioxide, and nutrients are exchanged in the body.

Blood's color is variable, due to its changing contents, especially oxygen and carbon dioxide. When the level of oxygen is high, blood's color is bright red, and when the level of carbon dioxide is high, blood's color is darker red. It's never blue! But in the diagrams, sometimes blue is used to show blood high in carbon dioxide, on its way from cells to the lungs to trade that CO_2 in for some fresh oxygen. Check out the path of blood through the whole body in this diagram:

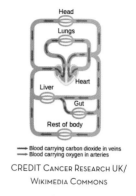

CREDIT CANCER RESEARCH UK/ WIKIMEDIA COMMONS

Now, let's go back to the model. Before you read on, try to work out which of the parts mentioned above and shown in the diagram is represented by which of the parts in your model.

Here's a table comparing the model and the real thing:

REAL HEART	MODEL HEART
4 chambers	2 chambers
Ventricle	Soap bottle
Atrium	Balloon
Arteries	Tube as it heads out of the soap bottle
Veins	Tube as it heads into the balloon
Capillaries	Not really represented
Blood	Soapy water
Mitral/Tricuspid valves	Coin in the bottle neck
Semilunar valve	Your finger squeezing the tube

The coin in the bottle's neck is a more accurate model of a valve than you squeezing the tube, because it's automatic: the inflow of the water from the atrium pushes it out of the way, but then when the pressure builds up in the ventricle, the coin slams back into position, stopping up the hole out the neck. This kind of automatic one-way valve is in all sorts of pumps, mechanical and biological. Most often there are two of these valves on either side of the pumping chamber, just like the two around the ventricle of the heart.

In reality, what you've made is a model of a fish heart. Fish have only one atrium and one ventricle in a row, along with some other funky fishy organs. This means the blood flow gets split off after the ventricle with some going to the gills, where blood can pick up oxygen, and some going to the body, where it can deliver the oxygen. This is less efficient than our system, but nonetheless seems to keep all the fish of the sea swimming soundly.

So your model is not complete. But if you were to make two of these models, you could hook them together into a full four-chamber heart. We did that at a training last year—check out the photo of Mestre Caetano and I pumping in unison. "Pulmão" is lungs in Portuguese and Tetun, and "Isin" is body in Tetun. So in this photo I've got the right ventricle and atrium and he's got the left ones.

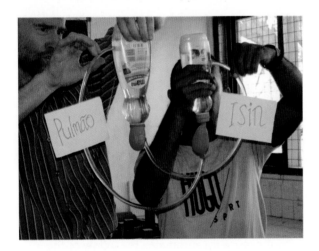

Now go get a real heart and check it out!

Gather more stuff

▸   Heart from a big mammal

▸   Tray or piece of cardboard

Gather more tools

▸   Retractable knife

▸   Optional: hose with water under pressure

▸   Magnifying glass, if you've got one

TINKER ON

Wash your heart off well before you begin—it's amazing how squeaky clean and almost synthetic it can feel. You can do this in a sink, but it's nice to have the freedom of a hose outside, so you can use a load of water and not worry about making a mess.

While you're washing, note the holes are all on the top. Depending on how the heart was cut out of the animal, there may be tubes connected to it. Try to figure out the four main paths, two in and two out. The two inbound should be by way of two elastic bags, which are the atria. The two out should be large, tough tubes, the pulmonary artery and the aorta, going to the lungs and the entire body, respectively.

In the photo here, at the top left Zaya is holding open the pulmonary artery, and just to the lower right of it is the aorta, the larger of the two. Those are where the blood comes out. In the lower left of the photo is the opening of the right atrium, and at the lower right Acede is holding open the left atrium.

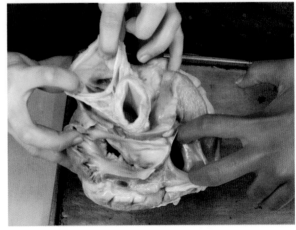

Here Zaya's finger is in the right atrium and her thumb is in the pulmonary artery, leading to the right ventricle. Blood went in with her finger and out with her thumb.

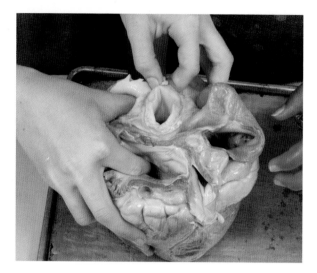

Here Acede's finger is in the left atrium and her thumb is in the aorta, coming from the left ventricle. There is another opening to the left atrium to the lower left of her finger. Again, the blood went in with her finger and out with her thumb.

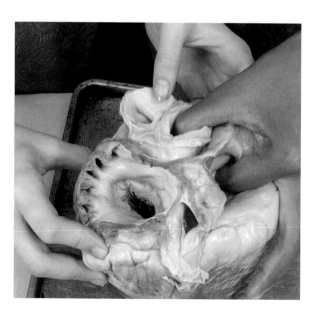

When you figure it out, you should be able to put the hose into an atrium and get the water gushing back out from the ventricle. Here the hose is in the right atrium, and the water is coming out the right ventricle through the pulmonary artery. This is the smaller outbound tube on top, heading for the lungs.

Now the hose is in the left atrium and the water is coming out the left ventricle through the aorta. That's the biggest outbound tube, soon to branch out and lead toward the whole body.

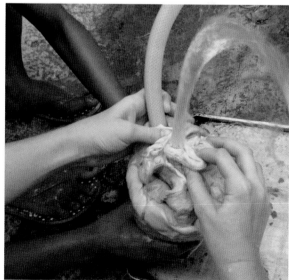

If you can't figure yours out by studying these photos, don't worry. The surefire clue to knowing which side is which is that the right ventricle is smaller and weaker. This makes sense because it has to pump blood only to the lungs, which are right next door. It makes up less than half the heart, maybe one-third, and is shown on the upper left in this photo. The lower right, and the entire base of the heart, is the massive left ventricle. (Left and right are according to how the organ sat in the beast's body.)

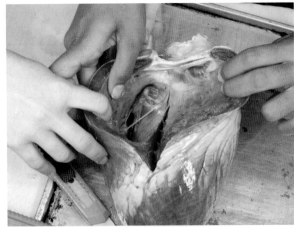

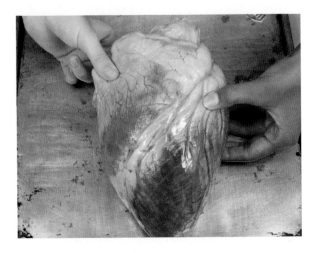

Find the ventricle's intake hole through the valves to the right atrium (finger to the left in this photo), and the outbound hole through the pulmonary artery (finger to the right in this photo).

When you think you've got the right ventricle identified, squeeze on it a bit and cut it open. It should be pliable and loose, whereas the left one will feel hard and stiff because it's made of much thicker muscles. When you find the inner cavity, cut around the edge to open it wide.

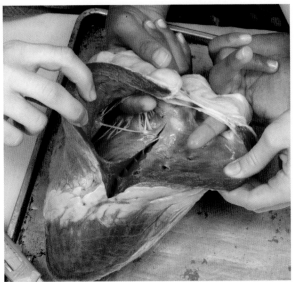

Check out those crazy white strings. Those are the hooked to the valves, where the heart is essentially pulling on itself to close a valve and pump.

Cut around a bit farther if you want, to open the entire right ventricle up for examination.

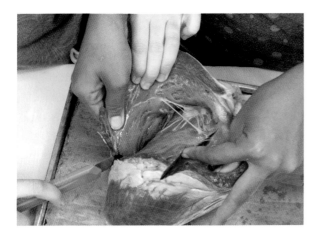

Now turn the heart over and start in on the left ventricle. It's going to be much thicker, so maybe you won't know where you are going until you find the inner cavity.

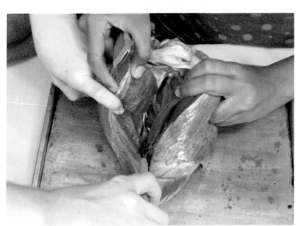

Try to see where the blood enters the ventricle and where it exits. Here Zaya's finger on the left is stuck through the entry hole—the left atrium—and the one on the right stuck through the exit hole—the aorta.

Remember, this ventricle is the strong one, so it makes sense it's got even more of those white cords linking the sides and the valve. Imagine the blood rushing in from the top, filling this cavity to the very bottom, and then surging back out the other hole as the giant muscle contracts.

Cut it open top to bottom to find the path of the blood in and out.

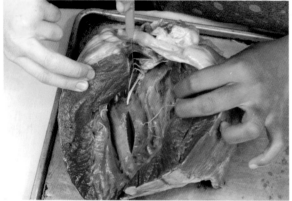

You can keep hacking around now, looking carefully at each part and trying to figure out how its characteristics make it work the way it does. Get out your magnifying glass and look closely at each part. Note the orientation of the sinews of the muscles; the connection points of the white cords; the thin, white fat layers on the outside; and the transition between soft red muscle and tough white vessel wall.

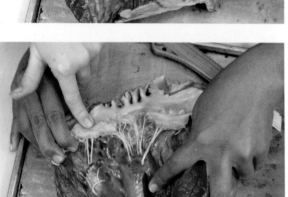

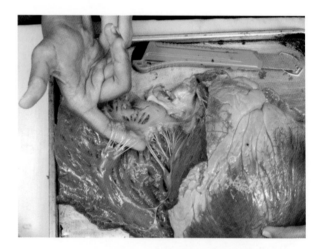

Notice how the characteristics of tissue and muscle change so drastically between the ventricle and the atrium and the aorta due to their completely different functions. Here the upper left is the atrium, the upper right is the aorta, and the whole bottom is the ventricle.

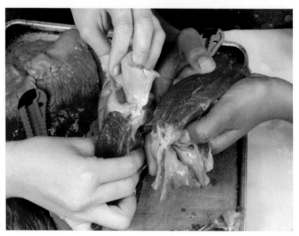

By the way, hearts are good eating, if you're a carnivore. They're full of iron and haven't got the body's waste enveloped in them like the kidneys and liver do. Here in Timor, they're slightly more expensive per pound than normal meat; we've never done this dissection without one of the teachers asking to take it home and cook it up for supper. One time I did a goat heart dissection with our entire after-school program on Valentine's Day and ended it with a barbeque of the results.

WHAT'S GOING ON?

I pointed out most of what's going on during the dissection. To wrap it up, think about how special this organ is. The heart's muscle is not like other muscles. You can't really control it, and thank heaven you don't have to think about it to make it go. Also, it's not pulling on a bone, like almost all other muscles. It just squeezes on itself, pushing blood out and then letting it flow back in.

The heart's function is critical for our lives. Many diseases can affect the heart, such as hypertension and excess cholesterol in the blood. These result in poor blood circulation and other complications. I hope it goes without saying that you should take good care of your heart so that it continues to pump well for you for decades to come.

Part IV
Earth and Sky

Chapter 14

SUN PATH MODEL

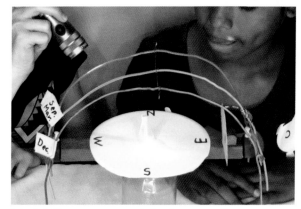

Model the seasons and shadows anytime, anywhere on Earth.

I'm interested in how modern amenities such as electricity and television affect people's observation of nature. I often ask students what they know about how the moon changes position and shape. (I'd ask you, but you're not here!) When I go to teach the phases of the moon, it's a challenge when my students have never made many observations of it.

Another slightly subtler question I ask is about where the sun rises. From your house, can you tell me where the sun rises and sets?

Well, shoot, you say, every school child knows the sun rises in the east, right? Well, sorta, but not really. Accurately speaking, it only rises exactly in the east two days of the year, on the equinoxes, which fall around the 21st of March and September. Other days it rises north of east or south of east. And the times change every day, nearly every place on the earth.[1]

If you didn't know that simple fact, ask yourself, why? Cave people six thousand years ago knew this and built giant astronomical monuments all around the world to show they knew it.

So now ask yourself: Did those folks have street lights? Did they watch TV until the wee hours and then sleep in on their holidays? Did they go to work and school far from their homes and spend most of their days indoors?

No. They spent their lives outdoors, close to home, and had only fire to light up their lives. When the sun came up, a day could really begin, so it was a heck of an event, one to be studied and understood as well as possible. Many of our ancestors even built religions around that fabulous ball of light and heat that comes around so dependably.

So consider yourself less informed than your prehistoric ancestors,[2] and also less informed than the average farmer in Timor, who like her ancestors, spends her life mostly outdoors and close to home, and has only recently got electricity. You better believe she knows where the sun rises. I've asked many farmers outside the capital, and not one has faltered in giving an answer. It's always a complex answer: now it's rising here, but in a few months it will rise there, and during the rainy season it rises over there, and so forth. Build this little gizmo to understand why that's true.

Gather stuff

▸ Stiff wires, three around 60 cm long. Pipe cleaners may work, though maybe you should braid three together and extend them. Here we show electrical wire, single strand (meaning one big copper wire inside, not many small ones, which is called "multistrand"). We stripped it out of the sheath with all three of the wires wrapped together, and it was nice that they were all different colors.

▸ Rectangular stick, around 24 cm long. Could be as fat as a 2×2 or 1×2 (sometimes called a "furring strip"), or any scrap piece that's sturdy and is flat on top and bottom (not a broomstick, in other words).

▸ Craft sticks, 2

▸ Pushpins, 4

▸ Paper plates, 2 or 3

▸ Stiff paper

▸ Small chunk of wood

▸ Twist tie

▸ Figure for the centerpiece

▸ Jar or can for base

▸ Flashlight, with a single bulb

Gather tools

▸ Wire cutter

▸ Wire stripper

▸ Hot-glue gun with glue sticks

▸ Pencil

▸ Marker

▸ Scissors

▸ Protractor

TINKER

Cut your stick to 24 inches.

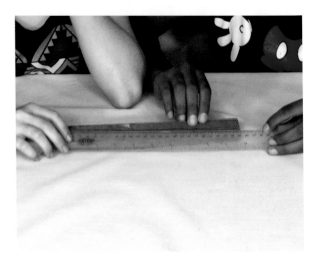

Now, a bit of geometry. We'll be making a semicircle, and building it around the base stick, which will span the diameter of that circle. Your mission, should you choose to accept it, is to calculate how long the semicircular circumference will be. That's how long your wires will need to be, plus a bit more on each end.

You may recall that the circumference C equals the diameter (d) times π.

$$C = d\pi$$

But we want only half of the circumference, so let's divide it all by 2:

$$\frac{C}{2} = \frac{d\pi}{2}$$

That fraction $\frac{C}{2}$ will be the length of our wires, so let's work it out. Our d is 24 cm, π is around 3.1, so π × d = 74.4 cm.[3] Then divide that by 2 and you get 37.2 cm. Round that off to 37 and then add 5 or so centimeters on either side and that's how much wire you need: 47 cm, more or less.

SCALING UP

With the previous equation, you can build this model at any scale you want. For example, if you want a big one, start with a stick one meter in length—100 cm—and you get this:

$$\frac{C}{2} = \frac{d\pi}{2} = \frac{100 * \pi}{2} = 157 cm$$

Add 10 cm on each end now, because it's bigger, and your wires should be 177 cm long. We built one like that, and instead of using wire, we used strips from a bamboo stalk.

The distance between the three strips should now be around one-fourth of the half circumference or 157 / 4 ≈ 39 cm, so the two side ones will be about 20 cm from the center one.

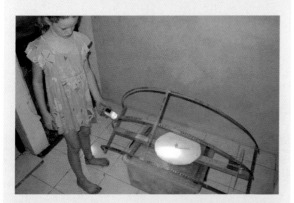

This is good for classroom demos. Zaya used it to show our neighbors how all this works.

Prepare three wires at least 47 cm long and lay them out on the table.

Find the midpoint, more or less, and twist another wire between the three so that the far two are about 9 cm apart. That means the two side wires are 4.5 cm from the center one.

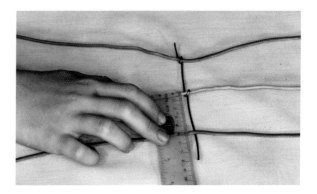

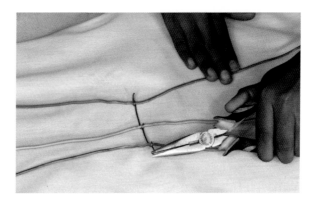

That 9 cm comes from another calculation, a bit harder to explain, so I'll do that in the "What's Going On?" section later. Trust me on it for now.

Now measure out from the center point half of 37 cm, which is about 18 cm. That's also three-fourths of the length of the stick, so we used that to measure it. Put some marks there on each of the wires.

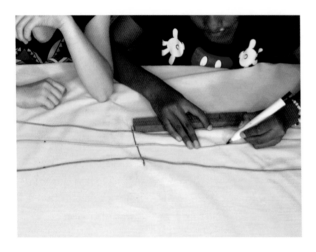

Lay down the craft sticks right at the marks you just made. Wrap the two outer wires around each craft stick, and leave the middle one just resting on top. Glue these wraps all into place, again 4.5 cm away from each other.

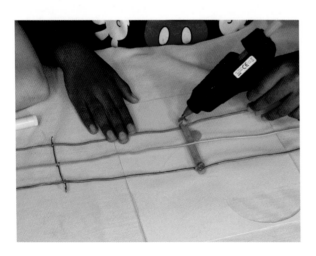

Put that aside and let's build the stick with rotating caps on either end. Put a pushpin through the center of each cap from the inside out, and push it into the center of the end of the wood. We had to use a hammer because the stick was tropical hardwood.

Now push another pushpin into the edge of the bottle caps. The purpose is to make a stronger glue joint.

Get the wires you prepared and chop them off evenly 5 cm or so from the sticks. Then bend the three into a semicircle around the wood stick so that the two craft sticks are vertical at the caps. Bend the middle wires out of the way so you can glue the caps to the craft sticks.

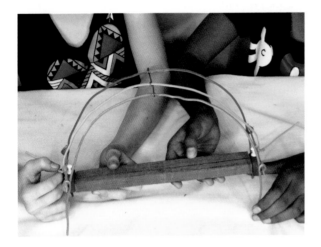

Now the touchy part: Without burning yourself with hot glue, glue the sticks to the caps. Get some glue stuck on those vertical pushpins for a solid joint, but keep it away from the wood and the central pushpins, which have to rotate freely. You have to hold everything steady for a minute or two while the hot glue cools and hardens.

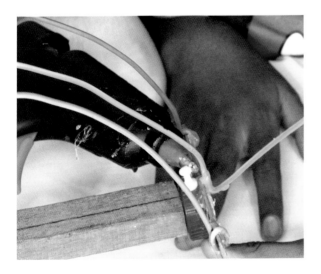

Glue the base stick onto the top of a jar and the main structure is finished. The wires should be able to tilt back and forth as the caps rotate on the pushpins. Cut the flat center out of two or three paper plates to make a sturdy base platform, and glue it on the center.

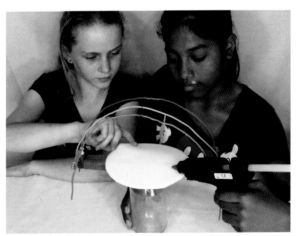

Write in the directions on the base platform: N, E, S, W, Never Eating Soggy Waffles, as you turn to the right, E and W on the ends with the base stick. The edge of this base represents the land you can see if you stand and look around. Anything below the edge is below the horizon and you wouldn't be able to see it.

Now it's time to make the angle dial. Get a little protractor (you should have one from Chapter 6). Put it on some card stock, like file folder paper, and trace around it. Mark the larger marks on the paper, every 10 degrees, but don't write in the degrees.

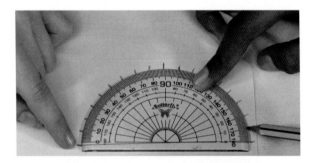

Cut it out and write in the degrees as shown in the following photo: 0 top center, counting down by 10s to the right and to the left. All the way to the bottom should be 90 degrees on both sides. Mark one side S for south and one side N for north.

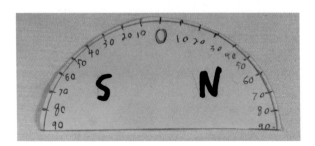

Now transfer the marks to the other side of the angle dial and write it all again.

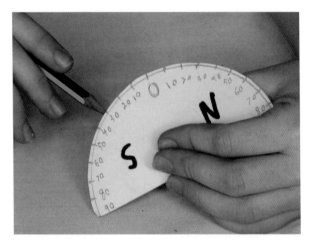

This is critical: keep the S and N on the same ends of the paper, as shown in the following photo of the back. This is so people can see the angle from either side of the dial.

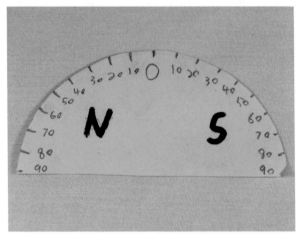

Now glue the little nub of wood onto the stick near the end, and glue this angle dial onto the nub. Important: Put your N to the left and your S to the right when viewed from that end, as shown. Note that it's *opposite* from the N and S on the base platform. This is how we have it, and though you could do it the opposite way, this will save confusion when you try to figure out the astronomy.

The final touch on the structure is the pointer for the angle dial. We have special bread here that comes with a golden twist tie, so that's what we use, though I suppose nongolden twist ties would work too. Twist it tightly around that center wire and leave one end out pointing to the angles. When the wires are straight up and down, it should be pointing to 0.

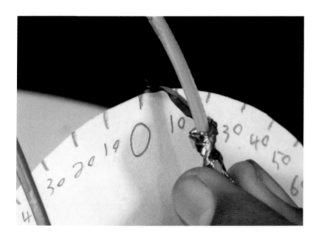

The angle shown on the angle dial indicates the latitude that you're modeling with the sun path model. When it's at 0, you're modeling a point on the equator. When you're at 40 degrees north, you're in the central United States. At 23 degrees south, you're in Rio de Janeiro, Brazil. And at 90 degrees, you're at one of the poles, North or South.

Now all we have left is to put something on the base platform and bring in the sun. We found four different things we could use for objects to cast shadows. Left to right we have a pushpin in the top of a bit of glue stick, a small machine screw, a LEGO Girl, and the top tip of a tooth-rotting ice pop with a yellow plastic BB stuck in the top.

We decided to use the ice-pop top because it's got a pointy top and a stable base.

The wires represent the path of the sun at different times of year. The blue on our model represents the sun's path at the winter solstice in December (brrrrr, in the Northern Hemisphere). Brown represents the sun's path at the summer solstice in June (hot, in the Northern Hemisphere). The bicolored yellow and green wire in the middle aptly represents the sun's path at the equinoxes in March and September. We put some tape tabs on there to label them.

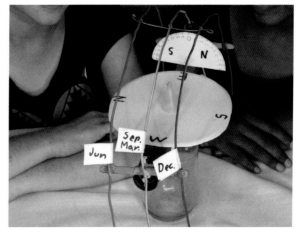

Now the stage is set. Let's turn the lights low and see what happens. (The real thing will be clearer than these photos, because I had to leave some lights on.)

Start with a location on the equator, say, Borneo. Make your wires stand straight up and down, dial pointing to 0 degrees latitude. We'll begin in mid-September, shining the flashlight along the central yellow-green wire. Bring the light up from the east, the magnificent orb rising to light the day, always shining directly on the ice-pop top.

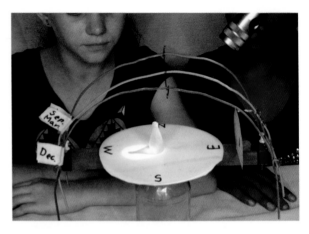

Keep moving it up to noon, afternoon, and evening, always watching the shadow.

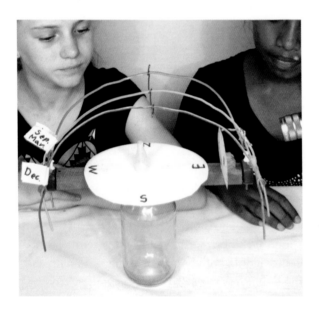

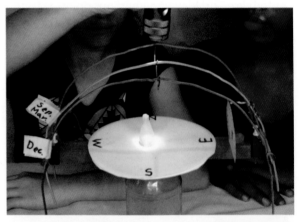

You should see a long shadow formed by the ice-pop top, though it's not clear in the image here. (You have to accept that a flashlight is a pretty pathetic representation of the sun in this model; it shines in only one direction and is far too close. That said, it can generate shadows that mimic the sun's, and that's our goal.)

Now begin moving the sun up in the sky. At about 10 o'clock you should see the shadow clearly on the white base.

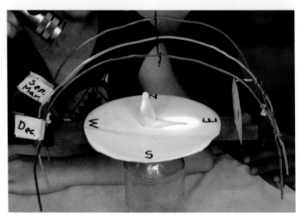

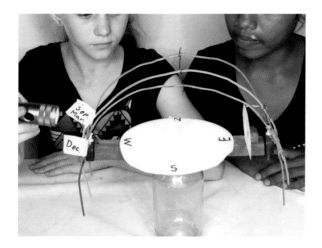

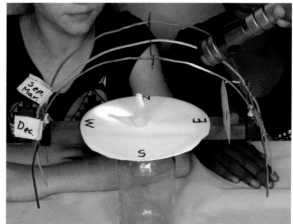

The ice-pop top shadow fades off to infinity as it enjoys a picturesque flashlight sunset. Did you see how at noontime there was no shadow at all? Ice-pop top was standing on its own shadow. That's the tropics for you.

Now we move to mid-December, just before Christmas, Hanukah, Kwanzaa, and innumerable pagan holidays. Something tells me this is an important time of year. Run the flashlight sun through the times of day, always shining the light along the line of the blue wire. Watch the shadow move.

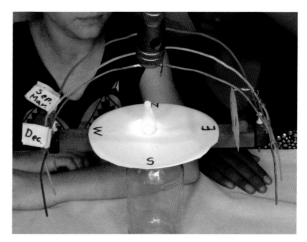

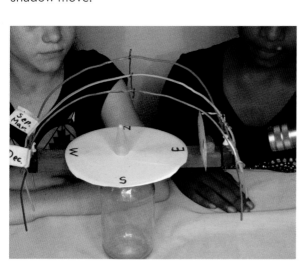

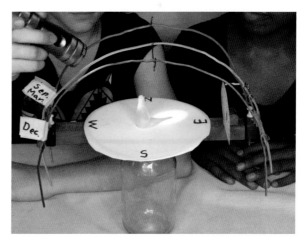

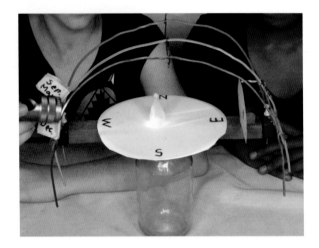

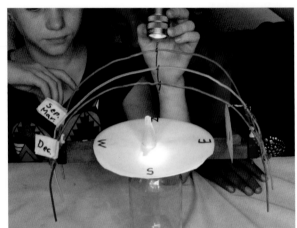

The key one to note there is noontime. Even though you're on the equator at noon, you've still got a shadow to the north. It's as big a noontime shadow as will ever be at this spot on the globe.

Now leap across the months to mid-June, the other solstice. Run the sun through a day on the equator, now shining the light along the line of the brown wire.

Notice what happened? The shadows are all reversed now. That noontime shadow still happens, but now it's to the south. Are you starting to see how this works? That's exactly what happens for folks living on the equator, in Sumatra, Congo, and of course, Ecuador.

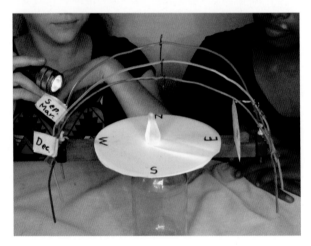

Now let's change latitude. I'll use my hometown, Maryville, Missouri, at around 40 degrees north latitude. Tilt the wires until the dial shows that angle.

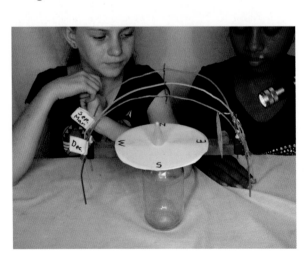

Now run through a day at each season again. Here's noontime in March:

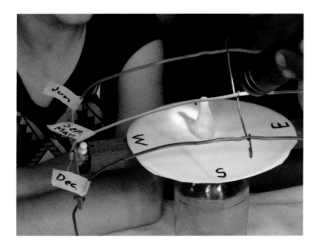

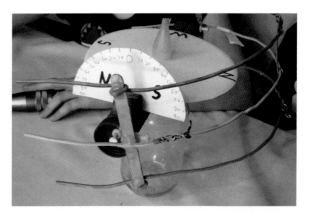

Check out that shadow! Jump down to noon in December and it gets even longer. Jump up to noon in June and you've still got a small shadow at noon, the shortest you'll get at this latitude. We're not in the tropics anymore!

Keep trying to see this from the point of view of your little ice-pop top. It's standing there looking up at the sun, and the wires are the path that the sun travels in the sky. It will never, ever see the sun straight overhead—a position called zenith—as long as it sticks around Maryville, Missouri. To see that, it will need to do some tropical traveling.

You can set this thing to any latitude and hold the flashlight at any time of day, any season, and see the shadows that will occur at those places and times. Let's try the South Pole! Turn the wires on their sides so that the dial is pointing to 90 degrees south latitude.

Here's 9:00 in the morning, December 21, South Pole:

Here's 6:00 in the evening, same day.

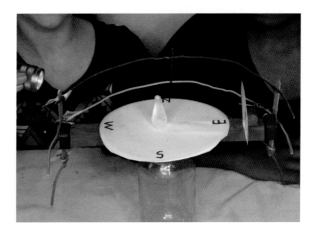

Same shadow, different direction! The sun is exactly as far above the horizon as it was at 9:00! You have to imagine the wires going all the way around, but you can show that at midnight the sun will still be there, shining away.

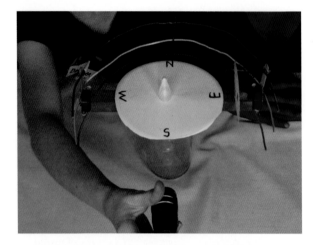

So the sun never sets in December at the South Pole; it just goes around and around the horizon. But now check out June, month of baseball and swimming, picnics, and suntans back in the good old USA. Here's 6:00 in the morning and 6:00 in the evening:

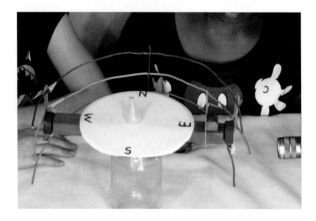

It's mighty dark there at the South Pole. The sun never even peeps above the horizon. If you want to go on a picnic, better bundle up and bring

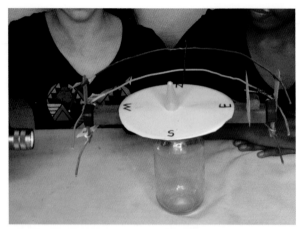

your flashlight. Bring an axe to chop a hole in the ice if you want to do some swimming. It's dark for the better part of three months at the poles while the sun is shining steadily on the other one. Ask your Aunt Wikipedia about polar night.

WHAT'S GOING ON?

Actually, I already explained more or less what's going on here. Following is more info on some questions you may have had.

How is the tilt of the earth represented in this model? Well, it's not represented directly. You need to have the arrangement in the following diagram clear before you can understand the model you just built:

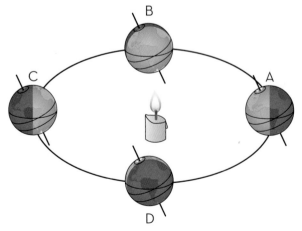

This illustration shows the earth always tilted to one side at 23.5 degrees as it goes around the candle—I mean the sun. (By the way, this is a great activity to do with an orange on a stick and a flashlight or candle in a dark room.) When it's on the right in this picture, the top of the globe gets more direct light, and thus more heat. When it's on the left in the picture, the bottom gets more heat.

If we take *up* to be *north* (that's just one way to look at it), then the globe on the right side portrays June, when it's warm in the north, cold in the south. The globe on the left side portrays December, when the opposite is true. The others portray September and March, when the sun's direct rays hit the equator. (Don't forget that it's spinning around daily so that it gets warmed evenly all round, like a rotisserie chicken.) None of that is in the model you just made, but the wires in three different positions are a direct result of this reality.

What's with that circular platform base? It represents the area that we can see when we stand on our patch of Earth and look around. On flat ground you can see a distant house maybe 20 miles away. Mountains will sometimes be visible more than 100 miles away, but after 20 miles the curvature of the earth makes small things like houses disappear. On the scale of a normal-sized globe, that's a circle the size of a pencil dot. So when you're holding a globe, you're looking at something truly grand in scale. This sun path model does not represent the whole globe, but rather just the spot where you're standing and the sun's path as it swings overhead every day.

If you don't have the pleasure of learning from a sun path model, you have to try to figure all this out by looking at diagrams like the following. They are based on the same circle of Earth that a person can personally view. This first one represents a location on the equator.

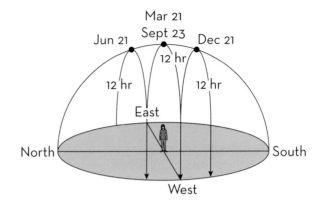

This next one shows the situation at around 45 north latitude, the latitude of Portland, Oregon, or Venice, Italy, or Harbin, China.

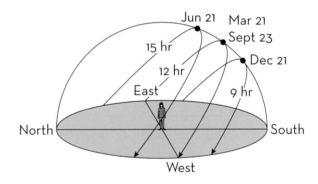

And finally, the situation at the North Pole. Notice how the directions disappear; at the North Pole, everywhere you look is south. The earth spins beneath you as the day goes by. The sun doesn't rise at all for months at a time, and then it doesn't set for months more.

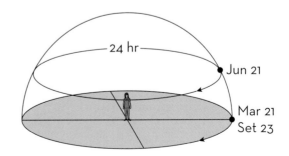

Are there seasons on every planet? Some planets have none! If the tilt is zero or next to zero, the planet's inhabitants will see the sun rise in the same place, every day, all year long. Boooooorrrring. This also makes for less interesting weather, as the daily angle of the sun's radiation in each place never changes.

Why 23.5 degrees? That's simply the fate of our chunk of cosmic dirt. Our planet and our solar system, along with all the others, was formed by the condensing and sticking together of vast swaths of star dust, left over from a star's death explosion, a nova. This dust and rubble was pulled together by gravity. We think of the earth's gravity pulling everything down, but actually gravity pulls everything together, us to the earth and the earth to us. So as the earth formed, all the rocks and dust swirling nearby were drawn together, smashing and sticking and heating up with pressure and friction. The resulting speed of rotation and tilt of the planet were completely arbitrary, determined by millennia of collisions. Other planets have other angles, determined in the same arbitrary manner.

Another key result of this angle is that anywhere within 23.5 degrees of the equator will get the sun directly overhead at certain times a year. Here in Timor, we're 9 degrees south latitude, and we get solar zenith around October 10 and March 1. We had

a party one year and stood around a giant vertical post and watched as its shadow disappeared.

How does 23.5 degrees translate to the distance between the two outer wires in the model (9 cm in the model shown)? Here's another rough-and-ready, seat-of-the-pants calculation. As the earth goes around the sun, the point of sunrise moves 23.5 degrees north of east, and then 23.5 degrees south of east. Together that's just over 45 degrees, which is one-eighth of a full 360. In our model, the circumference of the paths was 74 cm, and one-eighth of that is around 9. So there.

To me, this 45-degree angle, one-eighth of a full 360-degree circle, is one of the more fascinating aspects of this activity. One half of the sky's total circumference is the distance around the horizon from east to west. Half of that is from east to north, and a quarter is half of that. That final distance is 45 degrees, and it's how far the sunrise position moves between December to June. It's an incredibly large angle. Check out these photos I took from the same spot on the beach near our house:

Looking directly East from Bebonuk Beach, the moment of sunrise:

June 21

September 21

December 21

Pretty amazing, eh? I mean the picturesque beach is nice too, but look at how far that sun moved from June to December! Hard to believe so many people never notice it. Kill your TV!

Do this yourself! Get up early every month or so and check out where the sun rises from your front door.[4] When you've got it nailed down, put up some big rocks in line with the sun's rays and maybe people will still be pondering it in five thousand years. That's what happened at Stonehenge!

March 21

CREDIT: WIKIMEDIA COMMONS

ENDNOTES

1. Smack on the equator, the sun rises more or less the same time all year around. Near the poles sometimes it never rises, and other times it never sets. More on that later.

2. There is also the issue of horizon view. Many times in a city or a jungle or a deep mountain valley you can't see the sun rise. But there is generally a time when it comes into view, and if you check it every day from the same observation place, you'll find it moves.

3. You may protest that π is actually 3.14159, to which I'll gently reply: put that in the equation and instead of 74.4 you get 75.39—less than a centimeter difference. So it doesn't matter much, which is why I didn't bother with it. In most tinkering, the rough-and-ready round number is good enough. If you want to polish it up to professional standards later, you can always add more decimal places.

4. What about the moon? Check it out yourself! It also changes places, but in a slightly different pattern than the sun, and its cycle is only a month, so you need less patience to follow it.

Chapter 15

GLOBAL WARMING MODELS

Make models to understand why the earth is warming.

As I sit writing this in 2018, I've just lived through the earth's 18 hottest years on instrumental record with one exception. That's not to say there were not hotter years back in prehistory. When the earth was just forming some 4.5 billion years ago, nonstop volcanoes and meteor showers made for some toasty summers. But the steady trend for the last century has been warming, and the rate of warming shot up beginning in the 1970s. What we're seeing now is the fastest temperature change ever measured directly.

A number for the temperature of the earth is a pretty marvelously complex quantity, if you stop to think about it. Scientists arrive at this number with thermometers in the air, land, and sea, as well as satellite readings. Then the entire global scientific community watches the process and helps to make it more accurate. The vast majority of scientists agree that human creation of greenhouse gases like CO_2 has skewed the natural balance of the earth's heating and cooling. (Petroleum companies and those that grow wealthy from burning fossil fuels are scrambling to downplay the human impact. It's always good to follow the money behind an argument.)

Timorese are not worried about inundation of their coastal cities and villages as the polar ice melts and the seas rise. The island of Timor itself happens to be rising around 10 cm per year! More on that in Chapter 17. Timorese are, however, extremely worried about the climate change already happening as a result of this warming. According to the climatologists working here, the island is experiencing a frighteningly rapid shift in growing seasons, less dependable rainfall, and more severe weather overall. Farmers are being advised to maximize flexibility and explore new crops that are more resilient to erratic weather patterns.

This can be hard advice to follow, especially for older farmers. After all, if it worked for Grandma and it's worked for decades, why won't it still work? But change is also apparent. I've spoken to several old-timers here who testify to things not being as predictable as they once were, and some are quite concerned about it. To explain what science understands about what's happening is no easy trick. Here I'll show models we've put in the curriculum to help understand the science behind global warming.

Gather stuff

- Three identical cups of water
- A bigger transparent plastic container that will cover the first one overturned
- Another one made from glass
- A thermometer that will go from the outside air temperature to around 100°F , if you can find one

TINKER

The setup here is pretty simple: put the three cups of water out in the sun, and then cover one with the plastic cup or container and one with the glass, and leave one alone.

Actually, wait a minute: first measure the temperature of the water. Use your fingers, and also a thermometer if you have one. All the cups should be the same temperature; if not, you should figure out why not. Measure the temperature of the air as well. Write these data down so you can compare them if you do this experiment on another day.

Then cover the two up again and wait a half hour or more.

Feel the temperature of the outside of the containers. Uncover the two covered cups. Measure the temperatures again with your fingers and the thermometer if you have one, and write down these observations.

WHAT'S GOING ON?

Here are the results of our experiment (we'll use degrees Celsius because that's what people use here):

AIR TEMP: 31°C	TEMP AT START	TEMP AFTER ONE HOUR	CHANGE IN TEMP
Uncovered cup	30°C	40°C	+10°C
Cup covered with glass container	30°C	44°C	+14°C
Cup covered with plastic container	30°C	46°C	+16°C

Big news: The sun warms things up when it shines on them. Thank your lucky stars this is true, for it would be quite unpleasant living, say, on Pluto, where the sun does shine but with an intensity 1,500 times less than it does here. Shoot, even on Mercury, the closest planet to the sun, the temperature goes down to −183 degrees Celsius on the side away from the sun. Kinda gives new meaning to the phrase "You can stick that where the sun don't shine!" Brrr.

But it's more complicated than the sun shining on things and warming them up. The earth is covered with a layer of air that is mostly, but not entirely, transparent. The larger, upturned containers in this activity represent that atmosphere, and if your experiment turned out like ours, the covered cups warmed up a lot more than the uncovered one. Here in the tropics, we didn't need the thermometer to get to that conclusion; the original digital method of finger testing did the trick.

This makes sense when you think about it. The uncovered cup was getting the same heat

radiation as the covered ones, but as it got hot, any little breeze that came by carried the heat away from it. There was also some evaporation that took away heat.[1] The containers covering the other cups prevented air from circulating around their cups, so the air couldn't carry the heat away. That's one reason the covered cups got hotter. That concept is similar to the reason the earth doesn't get super cold at night: the atmosphere holds the heat of the earth down instead of letting it all radiate away like it does on planet Mercury, which has no atmosphere.

But another thing was happening with the cups that is not so easy to visualize. Radiation[2] from the sun was hitting those three cups, heating them up, and then some of it was reflected back out away from the cup. Well, on the uncovered cup, the radiation just took off back up into the atmosphere. But on the covered cups, some of that radiation was caught and kept by the walls of the containers. Some got back out through the walls—they are transparent, after all—but some got trapped. The containers themselves became hot, as you may have felt.

This is has a close parallel to what happens on Earth. The atmosphere is like a giant container covering the earth. Though it's not made of plastic or glass, some of its molecules act the same way; heat radiated from the earth and the warm air in the lower atmosphere hits these molecules and gets absorbed. The result is that those molecules are heated up.

In addition to preventing the sun's radiation from escaping back out, the glass container stopped more of the radiation coming in, more so than the plastic container did. I think that's why we got a lower temperature on the cup under the glass container.

Before I go on, let's do the next activity.

Gather more stuff

- 20 rocks about the size of tennis balls
- 10 marbles
- Chalk or a rock that makes marks
- A flat, blank wall near a blank section of ground, concrete, tile, or something more or less flat

TINKER

First, make a table for the data you'll collect during this activity. Here's the template:

ROCKS IN THE RECTANGLE	MARBLES THAT STAYED INSIDE THE LINE
0	
5	
10	
15	
20	

Mark off an area about 1.5 meters long and three-fourths of a meter wide, with one side as the wall.

Stand back maybe 2 or 3 meters and shoot marbles at the wall behind the rocks with the goal of getting them to bounce back out across the line. (If you're not used to shooting marbles, spend some time practicing your thumb and finger action.) Shoot them low on the wall, not more than 10 cm from the ground. Shoot all 10, and then count how many stayed inside of the rectangle. If a marble goes out one of the sides of the rectangle, it counts as staying in. Only the ones that come back out the top of the rectangle (the side parallel with the wall, closest to the shooter) should not be counted.

Record this number in your data table.

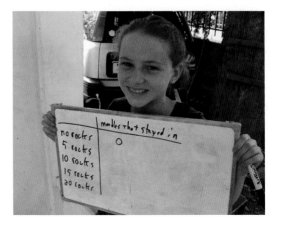

Now set five rocks at arbitrary spaces throughout the rectangle. Have the same person shoot 10 marbles again from the same position. (If a different person shoots them, his or her shooting skill may be better or worse than that of the first person, which would skew the results of the experiment.)

Record in your table the number of marbles that escaped.

Now add five more rocks spread out throughout the rectangle and repeat the activity. Then add five more two more times and repeat the shooting and recording so that the last time you've got all 20 rocks in there.

Now check out your data; here are ours:

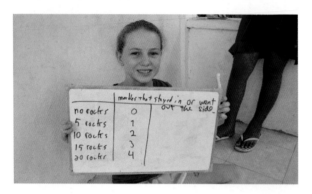

You can even make a bar graph of the data; here's the template:

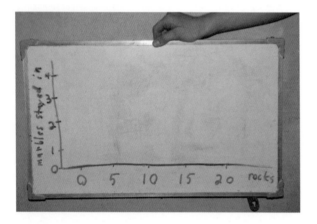

WHAT'S GOING ON?

Here's our graph:

I swear we didn't make this up. Our marble shooter, Binoy, is pretty deadeye and found little cracks to shoot through to get most of the marbles out, but each time we added rocks, one more marble got stuck inside or diverted out an end.

The trend is pretty clear, eh? The more rocks in the rectangle, the less marbles get out. Now what the heck does this have to do with global warming?

Plenty. This little game nicely models the sun's radiation hitting the earth. Here are representations in the model:

Sun: You shooting marbles

Sun's radiation: The marbles

Earth's surface: The wall

Atmosphere: The rectangle

Greenhouse gas molecules: The rocks

Get it? Those marbles that hit the wall are like the radiation from the sun that hits the earth. Most of it makes it to the ground and heats up the earth and water and forests. Then those hot things heat up the air and the air moves up and then radiates some of the heat back out to space. This complex process is represented by the bouncing up of the marbles.

Well, the ones that make it back out into space, heading off to who knows where, certainly don't do any more heating of our earth.[3] But others of these photons on the way up happen to hit various molecules in the upper atmosphere and stop dead their tracks, giving their energy to warm those molecules and contributing to the overall temperature of the earth. (Sometimes this happens with photons on the way in, too.) Those molecules, called greenhouse gases, act like the rocks in our game. The more there are, the more marbles (photons) get stuck, and the hotter the earth grows.

Around 250 years ago the most disruptive species on Earth, us, began to chop trees and mine coal to burn in order to power various machines that mass-produced stuff or transported people or stuff. It was called the Industrial Revolution, and the lives of many of us humans changed forever. The atmosphere also began to change. All that burning put more carbon dioxide into the atmosphere, and CO_2 happens to be a powerful greenhouse gas.

Well, we humans have never stopped, and in fact we've reproduced quite rapidly, multiplying our population by 10 times since the start of the industrial age. We also found all sorts of other things to burn, such as natural gas and oil. So now there are many more machines creating CO_2. The earth has various ways to take that CO_2 back out of the atmosphere, but they are limited, so it is piling up, together with other greenhouse gases, and the earth continues to warm.

This is not some vague theory. Scientists can dig out ice cores from Greenland and other massive ice shelves and check how much CO_2 was in the air at various times in the past, and make pretty good estimates of what the temperature was back then. From these sorts of data, it's abundantly clear that the CO_2 has been ramping up and the temperature has been rising in lockstep.

So what do we do about it? Fortunately, there is a practical answer: work as fast as possible for a conversion to carbon-free energy—that is, getting energy from sources that do not produce carbon dioxide as a product. Technology has made this 100 percent possible, right now in 2018, in the areas of electricity and factory production. Solar, wind, water, tidal, and nuclear sources *could* power all our stationary energy needs, right now, if we made that a national policy priority.

SCIENCE AND GLOBAL WARMING

By the way, there is always some wag, likely profiting in some way from burning petroleum, who says, "Dude! I remember some hot summers way back when I's a kid! That wasn't global warming! And think how cold it was last winter! I think we got nothing to worry about!"

To which the patient scientist will respond: "I always try to draw conclusions based on many data points, millions if possible, from all around the world. I also take a dispassionate—that means not particularly excited—view of any single data point, even if I experienced it myself.

If you agree, call up your representatives, local and national, and tell them this now. It doesn't cost a cent to make your well-informed voice heard, and exercising this right feels good to every patriotic American. This is the single most important step you can take to stop global warming and climate change that is even now causing problems for millions of already poverty-stricken people.

Harder will be the conversion of mobile petroleum use: cars, trucks, trains, boats, and airplanes. That liquid petroleum is just so frigging convenient! Imagine squirting a few liters of a certain *liquid* into your car's tank and using it to take you *effortlessly* 300 miles down the road. Unbelievable, eh? Like some kind of fabulous fantasy.

Unfortunately, aside from being unbelievable, it's also darn near unstoppable. This is one of the most critical problems of our times, and it will take more technology and creative energy policy to make this conversion of mobile energy use. Batteries will be part of the solution, but not all. Still, it should be a top priority. I encourage you to devote your life to solving this problem, with both innovative technology and creative public policy. Then, when you've got the job done, if you happen to think of it, give a small word of credit for the one who inspired you...

It's a mistake to say that global warming is ruining the earth. The earth's biospheres will adapt and carry on in the event of a huge temperature rise. Don't forget that a lot of plants love global warming! Turn it up! Bring on the radiation! Unfortunately, the changes will spell disaster for many habitats we humans have come to depend on, and the poorest will suffer the worst, and the first. So my family and I are doing our best to use less carbon and green the earth in order to curb global warming.

ENERGY CONSERVATION IN THE MAJORITY WORLD

A few years back, our group of science teacher-trainers first began to teach teachers and students here in Timor about global warming and climate change. When we came to the part about using less energy, cutting back on car trips, shutting off lights at night, I started to feel a bit silly. Many Timorese just got electricity to their homes in 2013, when the government carried out a big expansion of the national grid. Most are paying for this electricity, and even though it's subsidized, it's a significant cost to their household, and so on the whole they're incredibly conservative: dim light bulbs, shutting things off right away, etc. And as far as transportation, pretty much everyone rides motorcycles or takes public transport anyway, so it's not like there is much fat to cut away there either.

Even traditional houses made from palm branches and bamboo get electric meters to count and pay for the energy used.

So most Timorese need to use *more* energy, not less, to attain what we in the United States would call a lower-middle-class standard of living. Sure, they should avoid wasting energy, but good grief, if I'm going to use my time and energy to counsel people not to waste energy, I'm not going to do much of it here. I'm going to go back home and talk to my neighbors who put high-intensity spotlights on their Christmas decorations and drive to the mall in an SUV!

What we decided to focus on instead in our environmental curriculum here is the importance of forests in the fight against climate change. Trees prevent erosion and maintain productive topsoil as well as clean the air and make oxygen. They're also beautiful and, when managed right, can produce other valuable things. Taking good care of Timor's forests is something the Timorese can do that benefits themselves and the whole world too.

ENDNOTES

1. If you want to see the effect of evaporation, compare two cups of water without covering them, but put a thin layer of oil atop the water in one of them. In the cup with oil on top, you'll get much less evaporation, since oil doesn't do it as readily as water We got a 3-degree difference in temperature, showing evaporation's role in getting rid of the heat.

2. Radiation is technically any energy that can be radiated. The sun gives off radiation at many different frequencies, from X-rays to radio waves, all of which are electromagnetic radiation. Visible light and infrared are there in the middle of the electromagnetic radiation spectrum, and thus are also called radiation in physics. The kind of radiation you have to worry about is called ionizing radiation, because it can mess up the molecules in your cells. This requires higher energy and begins around the frequency of ultraviolet. Wear your sunscreen!

3. The atmosphere is like a thin skin of air stuck to the surface of the earth. It has no "sides" to speak of: if you were to walk forward forever, you'll just go around and around the earth. Marbles bouncing out the side in our model are heading off more or less parallel to the ground, so they're not heading back out to space; thus we can count them as captured.

Chapter 16

ROCK CYCLE

Simulate one path in this complex web.

The water cycle is fairly straightforward: evaporation, condensation, and precipitation. Add transpiration—water getting sucked up by plants and going back directly into the air—and a few complexities of groundwater, and that's the whole thing! The rock cycle is another animal altogether. Not only is it complex with multiple interlinking paths, but it proceeds in such slow motion that we can hardly see anything actually happen in the course of our lifetimes.

That said, it's pretty important to learn if you're to understand the vast diversity of rocks you'll find when you begin observing the ground around you. Each one of them was formed through a certain path within this cycle. This activity will show one path and illuminate the others.

Gather stuff

- Two different-colored candles
- A few crayons of different colors
- Plates
- Matches or a stove
- An aluminum can or two

Gather tools

- Knife or scissors
- Spoon

TINKER

First shave off a little pile of paraffin from both candles.

Then shave off some of the crayons.

Next, carefully dig out a section of it. Try to get a biggish chunk without it falling apart. When you've got it, put it gently on a plate. There is your first simulated rock sample.

Dump it all together, and then fill up the base divot of an aluminum can.[1]

Smash the shavings down as firmly as you can with the spoon. Add a bit more and smash it all down again.

Now prepare a can for heating. It could be the same can, but we used another. Cut it around the base, but leave a little section so that it's still hooked on.

Fill up the divot again with candle and crayon shavings, including the remains of the first operation.

Smash this down as well as you can with the spoon.

Light a candle (on a nonflammable surface) and hold the can bottom full of shavings over the flame for 6 seconds only. You should see the paraffin start to melt around the edges and the bottom.

SAFETY NOTE Get an adult to help you set up and light the candle. Have a fire extinguisher ready. You can do this part on a stove as well, instead of the candle.

Take this off and let it cool. You can smash it down a bit more with your spoon after it cools. Again, carefully dig out a section of it and put it on the plate with the other rock sample.

Now put all remaining shavings into that same can bottom, pack it down with the spoon, and heat it again over the candle flame. This time, continue to heat it until it's all melted to liquid.

SAFETY NOTE Be super careful here; this liquid would cause a major burn if you spilled it on yourself.

Put it aside and don't bump it until it's cool, which could be as long as 5 minutes, even with a fan. Make sure it's all nice and cool before you dig out the last piece. Put it on the plate with the other two rock samples.

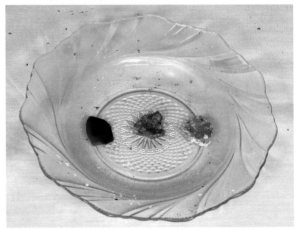

WHAT'S GOING ON?

What you've just witnessed is one path on the rock cycle. Here is a schematic of the whole cycle:

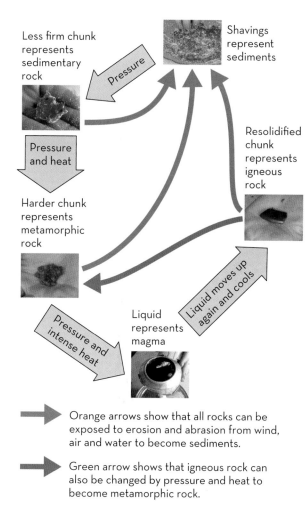

Less firm chunk represents sedimentary rock

Pressure

Shavings represent sediments

Pressure and heat

Resolidified chunk represents igneous rock

Harder chunk represents metamorphic rock

Liquid moves up again and cools

Pressure and intense heat

Liquid represents magma

Orange arrows show that all rocks can be exposed to erosion and abrasion from wind, air and water to become sediments.

Green arrow shows that igneous rock can also be changed by pressure and heat to become metamorphic rock.

pushed farther down, it gets more pressure and a lot of heat and its structure is altered. This is called metamorphic rock. Then if it keeps on going down, it can't take the heat and it melts into magma. But when magma gets thrust up back into the lithosphere, it cools to become igneous rock again.

One could say you also represented erosion and abrasion with your knife when you shaved the candles and the crayons. This is more or less correct, but since each candle and crayon represents a certain mineral, individually the candles would represent a chunk of that pure mineral, and this just doesn't occur much. The vast majority of rocks are complex combinations of minerals. When they erode, they end up as a complex sediment.

Notice that not all paths are possible in this cycle and not all arrows go both ways. For example, you don't get magma straight from sediments. They have to get smashed and mashed deep into the earth to reach the heat necessary to melt them. Also, metamorphic rocks can get busted apart with erosion, but when the sediments come together again, it's sedimentary, not metamorphic yet.

Of course, this model is a vast generalization. Reality is always more complex. A major sedimentary rock called limestone does not come from erosion at all, but rather millions of years' worth of dead organisms with shells, big and microscopic, falling and piling up on the floor of a lake or ocean, sometimes on top of a coral reef. And you can find limestone that is harder than some metamorphic rocks. For that matter, the igneous rock pumice, produced as the froth of volcanoes, is so soft you can scratch it with your fingernail.

Still, scientists use models precisely because they are simplified versions of a complex reality, and many parts of this model are right on. Many igneous rocks are more solid than metamorphic ones, many sedimentary rocks are soft to

Look at that gaggle of rocks and arrows! First off, let's name the five stops around the edge. Rocks can be divided into three categories: sedimentary, metamorphic, and igneous. The two other stops are the sediments themselves and molten rock, called magma.

The path you trod was from sediment to sedimentary to metamorphic to magma to igneous. You began by preparing sediments from several minerals: two main minerals—the candles—and other trace minerals—the crayons. Simple pressure on the sediments due to other layers being laid down on top leads to the formation of sedimentary rock. Then if that particular chunk gets crumbly, and the processes you did to get the model samples are all representative of what happens to real rocks. The model is your friend!

If you'd like to tinker more with this, I suggest going into the nuances of metamorphic rocks. You could prepare different kinds of sediments, some with very little red, some with mostly red, some with a lot of a certain kind of crayon. Then when you smash and heat for 6 seconds, see what you find different and similar between the samples. There are many types of metamorphic rocks, all depending on the composition and the processes that created them. Try to make as many as you can!

ENDNOTE

1. I've seen this done in the United States by wrapping the crayons and candle shavings in aluminum foil. We can't get crayons or aluminum foil so easily here, so we developed this one with the ever-present aluminum can, and I think it works a lot better.

Chapter 17

PLATE TECTONICS

Fathom how the ground beneath us moves.

We named our kid after the tallest mountain in Timor: Ramelau. At 3,000 meters (10,000 feet), it's 800 meters taller than Australia's tallest mountain. The continent of Australia is enormous compared to the island of Timor—200 times the area—and has its own continental shelf under the sea. But still, Ramelau is not hard to understand if you believe in plate tectonics.

When I say "believe in," I'm not being silly or religious. I'm saying when scientists look at data that have been collected over decades of research, they can decide that one theory is more effective than the others at explaining the situation and thus probably correct. They believe it. The beliefs held by scientists are based on the interpretation of data and always open to revision by new data or new interpretation. We try our best to avoid being skewed by emotion.

When I was young in the 1970s, my teachers didn't teach me about plate tectonics. It was a brand-new idea and somewhat outrageous. After

all, aren't continents that emerge, float around, and sometimes sink beneath the ocean the stuff of science fiction? But it turns out that one of the issues that dogged the theory of evolution for so long was still dogging scientists in the mid-twentieth century: the gargantuan expanse of geologic history.

The earth is so old you can't imagine it. Think you can? Forget it. You'll be on the fortunate side of statistics if you live to be 100 years old. If you're really fortunate, you heard stories from your 100-year-old great-grandma before she died, and before you die you'll meet your great-granddaughter and repeat some of the stories your great-grandma told you. That's seven generations, and it hardly ever happens. A personal, once-removed story spanning five generations is about all you can usually get—less than 200 years.[1]

But the earth is 4,600,000,000 years old—twenty million times longer than those stories extend! And shoot, archeology finds the oldest bones of proto-humans are only 2 million years old, which is still only 1/2000th of the history of the earth. Check out the earth's history timeline later in this chapter.

The conclusion? We humans are super-newcomers to the planet, and as individuals we just don't stick around long enough to see many rocks change. Some tectonic plates are moving at the rate of around a centimeter a year. So in 200 years, you've got about two meters of movement. I don't care how meticulous Great-Grandma was in her records, you can't expect to see this difference with normal observations. The geologists of old can be forgiven for thinking mountains and valleys and continents have always been there and will never move.

But move they do, relentlessly, unstoppably, and with dire consequences for the lands we call home. In these tinkering activities, you'll use crackers and cookies as tectonic plates and watch them move. What you're seeing happen in your little models here takes hundreds of millions of years to happen on the surface of our planet.

Gather stuff and tools

▸ Plates, 2

▸ Water

▸ Bananas

▸ Fork

▸ Table knife

▸ Crackers (Stiff ones work better, stiffer than saltines, stiffer than graham crackers, though those kinds will also work.)

▸ Powdered sugar and/or sprinkles

TINKER

That looks like a fairly happy little set of tinkering supplies, eh? And it's true: if you're careful and keep everything clean, you can *eat* the results of this first tinkering session!

First we'll set up the asthenosphere, the upper plastic-like part of the mantle. This will be represented by smooshed-up bananas in your plate. You don't need a deep layer of banana mush to make the asthenosphere—maybe one or two normal bananas[2]—but be sure it's smooshed up well with the fork.

The square crackers will represent tectonic plates that float atop the asthenosphere. We'll investigate the three different types of interaction between plates:

▸ Convergent (pushing together)

▸ Divergent (pulling apart)

▸ Transverse (moving lateral to one another)

First the transverse: get two crackers and chisel shallow teeth into one side of each. The teeth should more or less fit into each other, though it's not important that they fit tightly tooth to tooth.

Before you drop them onto the banana asthenosphere, try the setup on the tabletop; put them together tightly and then push both together and laterally—that is, one forward and one backward.

They should not move yet—just put on pressure in these two directions. Now chuck them into the plate and set it up again on top of the mush.

Begin putting force on them like you just practiced, pushing them together as well as one forward and one backward. This time, continue pushing harder and harder until they suddenly move.

That was an earthquake, by the way, and a decent model of a transverse tectonic plate motion. Try it a few more times while you've got it set up.

The next one is divergent plate motion. This is the easy one. Put two crackers in, press them down a bit, and pull them apart.

Indonesian-made, a staple of our snack diet here) are quite resilient, and we leave them for almost a minute. Graham crackers will sog up in a few seconds.

When you have a somewhat soggy edge to your cracker, put it on the top of the banana mush and put another hard one next to it.

Now steadily press them toward each other and watch what happens.

Notice the mid-oceanic ridge that forms from the banana mush rising between the two crackers. More on that in a bit.

Finally, the convergent model. We'll do two versions. The first involves softening one of the crackers. Put some water in a plate and lay a cracker so that maybe one-third of it is in the water.

How long you leave it will depend on the kind of cracker. The ones shown here (Square Puff

You can see that on ours the soggy part crumpled up on top of the hard one; a mountain was raised.

We can model another way mountains are raised when plates converge. Here we'll use our sprinkles and/or powdered sugar. So lay two new hard crackers (no sogging necessary) on top of the banana asthenosphere and spread a generous layer of sugar or sprinkles on top of both of them.

When one is clearly going under the other, gently and steadily push them toward one another and witness again accelerated geologic history in your plate.

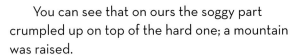

Now help one of the crackers go down below the other. This motion is called "subduction."

Did you get a magnificent little sugar mountain, like we did?

WHAT'S GOING ON?

Here is a diagram of the earth's tectonics plates. The key thing to notice is the borders between them, where the situations you just modeled take place.

CREDIT: UNITED STATES GEOLOGICAL SURVEY

Let's start with the transverse situation. In my adopted home of California, the San Andreas Fault runs the length of this state. The plate to the west of it is slowly scraping to the north, and the plate to the east is slowly scraping to the south. In 1989 a massive earthquake called the Loma Prieta struck just before game 3 of the World Series and was broadcast live to the world. This serious plate movement ruined a large swath of the greater San Francisco area, as well as my future home of Watsonville.

The Pinnacles is part of an extinct volcano made into a national monument (and great rock climbing spot). Most of the remains of this volcano are not far from Watsonville, but part of it is 195 miles away to the south! That massive old volcano erupted 23 million years ago right on the San Andreas Fault line, and the plates kept on moving, carrying one half south and one half north. Amazing.

Divergent plate boundaries are found in the middle of the Atlantic Ocean among other

places; many maps show this. The rocks found at these boundaries are straight from the asthenosphere, and the age of the rocks increases with the distance from the center of the plate boundary. This is evidence that the plates are moving apart and that new rock is formed when the hot ooze from the asthenosphere comes up in the resulting space, just like your smooshed-up banana did.

Converging plates often create mountains. In reality, there is no exact parallel between hard and soggy crackers. Plates are made of all different sorts of rocks, and whether one goes up over or down under depends on a bunch of different factors. We softened one cracker just to show more or less what happens when one bunches up in a converging situation.

To tell you the truth, a geologist friend of mine said the crackers in this model are not quite accurate, because real plates are flexible and bendable. She was talking about rocks like granite, basalt, and so forth—not exactly what I would usually describe as bendable. But when you look around Timor, you realize it's true; rocks bend all the time at great pressure and temperature, and this results in folds, buckles, uplifts, and tilts, and when they break, it turns into a fault. So the soggy cracker is actually quite an accurate representation.

Mount Ramelau, and the entire island of Timor-Leste, is in the other category of converging plates, the powdered sugar and sprinkles one. Timor Island is sitting on the far northern edge of the Indo-Australian plate, and this plate is being subducted—that is, shoved underneath the Eurasian plate. Logically speaking, you'd expect the land here to be going down and the land on the Eurasian plate to be rising and forming mountains as it goes up over us. But when you look at the rocks in Timor, you find a bunch of rubble: mostly bent, smashed, fragmented metamorphic and sedimentary rocks. This is the kind of stuff that's being scraped off the Indo-Australian plate and resulting in the mountains of Timor.

Due to this rubble piling up, Ramelau increases in height around 10 cm each year, along with the whole island. That's why folks here are not going to be affected much by the sea rising due to global warming; they're rising even faster!

The highest mountain in the world is Everest, and it just happens to *also* be on the border of the Indo-Australian plate and the Eurasian plate. The same motion is happening there, with the Indo-Australian Plate pushing north and the Eurasian plate pushing south, and that similar situation has resulted in the rapid (you know, several tens of millions of years) uplift of the Himalayan mountains. Pretty fabulous, eh?

So that's the basics of the motion of the tectonic plates. But there are no enormous fingers reaching down to push the real ones around, like you did in your model. Where does the force come from then? Tinker a bit more and the answer will emerge.

Gather more stuff

- Wok or big frying pan
- Water, 0.5 liters (two cups)
- Flour, ¼ cup
- Fork
- Rectangular sandwich cookies
- Stove

TINKER

Pour the water into the wok and stir in the flour. This is like making gravy, if you've ever done that, but with no grease.

> **SAFETY NOTE** Get your kitchen-friendly adult to help you heat up the stove.

Heat it on medium and observe what happens. That's the asthenosphere you're gazing at. Let it get nice and hot, maybe even boil a bit, until it turns a bit viscous—that is, sticky.

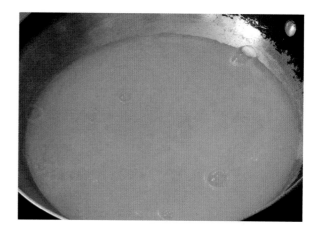

Can you see the little convection currents rising and swirling? That's what happens in the asthenosphere.

Now let's cover it with the lithosphere—the sandwich cookies. Turn the stove off for a moment. Lay the cookies on the surface of the gravy, wall to wall, right up next each other. Break some apart along diagonals to fill the open spaces around the edge. Leave some of them two layers, but take most of them apart to make one-layer pieces. The two-layer pieces will represent places where the lithosphere is thicker, like continents and mountain ranges. The one-layer pieces will represent thin plates and specifically oceanic plates. After you split one, if one side still has the disgusting sweet filling stuck to it, place it with that facing up. That stuff can represent the oceans.

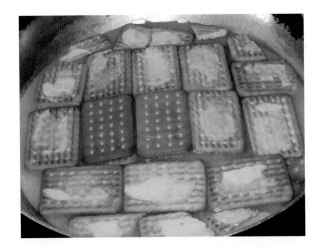

Now turn the stove back on low and feast your eyes.

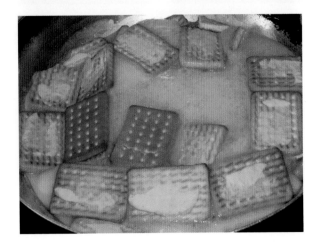

Everything is happening all at once here; you should be able to see examples of all three types of interaction among the tectonic plates in your gravy:

- *Diverging zones,* where the ooze is rising to create little mountain ridges between the edges of the plates, maybe in the midst of a frosting ocean

- *Converging zones,* where two plates are smashing into each other and one or both are tilting up, or one is sliding under the other

- *Transverse zones,* where two plates are scraping laterally to each other

It's a veritable mini-tectonic-plate-a-rama in a wok, all powered by the convection and bubbling of the gravy asthenosphere.

WHAT'S GOING ON?

The underlying (ha!) question has been answered: the force for moving the tectonic plates around comes from convection and circulation of the asthenosphere. This circulation was visible in the asthenosphere of your wok. The tectonic plate cookies were floating arbitrarily around on top of the churning asthenospheric gravy. That's just what the real ones do, with the following differences:

- They move slowly, remember? Centimeters per year.

- They are enormous; the Pacific plate is around 20 percent of the whole earth's surface.

- They are much denser and are formed of rocks instead of cookie material.

- For the most part, they're not edible.

It's important to see the position of the oceans. All this talk about floating may have you thinking that there is water underneath rocks.

In reality, the world's oceans are all sitting atop oceanic crust, which is in turn floating on the semi-liquid asthenosphere—your gravy in the model. That's three distinct layers at the oceans: asthenosphere, lithosphere, ocean water.

Now, your wok model actually had boiling going on, with bubbles coming up. This is not exactly like the asthenosphere, in which the magma is not boiling per se, but actually the role of water at great depths—and that means great pressures and great temperatures—is one of the fundamental factors in rock formation and volcanic activity. So your little wok model is not so far off.

One thing that often happens in the wok model that is not so realistic is that cookie plates pile up and spread far apart, leaving a bubbling area of gravy asthenosphere with no cookie lithosphere above it. As I mentioned earlier, when plates diverge the liquid material from below flows up to fill the space. But it cools pretty much immediately and forms new rock. It doesn't stay molten and bubbly very long. In fact, it's important to note that the lithosphere is composed of a lot of the same stuff as the asthenosphere, the two mixing and blending back and forth as rock materials melt at lower, hotter levels and solidify at higher, cooler levels.

As I write this, the Kilauea volcano in Hawaii is spewing forth lava, reminding us that there is hot liquid down there—and also that we should give thanks for a hard, cool surface up here to live on.

GEOLOGIC TIMELINE: CAN YOU FATHOM IT?

One way to get some grasp of the history of the earth is to graph it on a scaled timeline. Many timelines in books don't keep a single scale for the whole length of the earth's history—they smash the early years and stretch the recent years. You can understand that, because it's pretty darn difficult to figure out the details of what happened in the early days of the earth, 4.6 billion years ago. You don't need to leave as much space there, because there is nothing to write! ("Another meteor fell, another volcano erupted, ho-hum.")

But sometimes it's useful to put out the whole line so you can just gaze amazed at the grand expanse of time and the undeniable insignificance of our presence on Earth. For a scale you can use 1 meter = 1 billion years, and it *may* fit in your room. That's how we do it at our little geology and paleontology exhibit here in Dili, as you can see in the photo here.

Or you can make 5 meters = 1 billion years and fit it in a basketball court. You can hang it on a wall or a fence or mark it right on the floor. Here is a chart with some of the key things to put on your timeline.

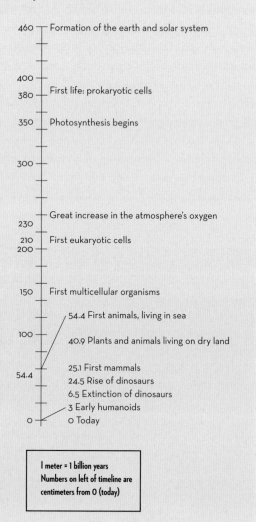

460 — Formation of the earth and solar system

400 —
380 — First life: prokaryotic cells

350 — Photosynthesis begins

300 —

230 — Great increase in the atmosphere's oxygen

210 — First eukaryotic cells
200 —

150 — First multicellular organisms

54.4 First animals, living in sea

100 — 40.9 Plants and animals living on dry land

54.4 — 25.1 First mammals
24.5 Rise of dinosaurs
6.5 Extinction of dinosaurs
3 Early humanoids
0 — 0 Today

I meter = 1 billion years
Numbers on left of timeline are
centimeters from 0 (today)

These milestones of life on Earth are the most interesting to me, but there are many other geologic milestones to put on as well—such as formation and breakup of the mega-continent Pangea, and major meteor and comet impacts. It's a great mini-research project your Uncle Google would be happy to help you with.

ENDNOTES

1. But please do your best: listen to your grandparents' earliest memories, ask them for stories their grandparents told them, and be sure to tell these stories to your children and your children's children. Actually, it doesn't even matter if you keep it in the family; anybody's grandparents and grandchildren will do. This is all part of the great ongoing experiment we humans are carrying out on Planet Earth.

2. I say "normal" for folks in the United States. In Timor, saying "normal banana" is like saying "normal hairstyle"—what the heck is that? Here there are usually at least five kinds of bananas for sale at any little fruit stand, and the number of banana types available if you shop around runs into the double digits. The flavor, texture, consistency, and color of each one is different. When is the last time you chose the flavor of your banana? The poverty of bananas in the U.S. is understandable, because shipping would knock out most of the varieties here: they bruise too easily, don't stay long in a pre-ripe stage, or are prone to bugs taking up residence. We try to eat bananas every day here to take full advantage of this local luxury.

Part V
Force Making Movement

Chapter 18

FLIP-FLOP AIR GUN

Make compressed air shoot stuff safely.

In this age of mass shootings, the last thing I think we need is more weaponry. That said, you can learn a heck of a lot from making this little beauty, and nobody will get hurt either. It's powered by compressed air and comes from a long tradition of artillery generally termed "spud guns." That's because they usually shoot sections of potato. Instead, we'll shoot sections of old flip-flops and paper rockets.

I have seen three types of spud gun: explosive, plunger, and air pump. The explosive one requires spraying highly flammable liquid into a chamber, closing it, and lighting it with some kind of spark. This is good fun, but it can also hurt you badly if you screw up.

The plunger model is the simplest, with a stick plunger pushing one spud section up the tube toward another section stuck near the other end. If the spud sections each seal off the tube, you can see how the pressure would increase as you decrease the volume between them. It's the science of Boyle's law: in an ideal gas, the volume and pressure are inversely proportional. (See Chapter 19 with Newton's slingshot activity for more info on proportionality.)

What we'll tinker with here is the final method, midway between the other two in complexity and in danger as well. If you tried really hard, you could probably put out your eye, so please try really hard not to.

Gather stuff

- PVC tube, ½ inch, around 2 meters, along with the following connectors, also ½ inch:
    - T
    - Ball valve
    - Cap
- PVC cement
- Epoxy glue, "5-minute," suitable for hard plastic
- Valve stem from bicycle or car tire
- Candle
- Matches or lighter

- ▶ Old flip-flop or piece of dense foam rubber, such as camping pads or packing materials
- ▶ Paper
- ▶ Tape, clear
- ▶ Stick or rod that fits into the ½-inch PVC
- ▶ Short stick to extend ball valve handle
- ▶ Hose clamps that fit around the valve handle, 2
- ▶ Soda bottle, 1 to 2 liters. Use only carbonated drink bottle here, because it was designed for a bit of pressure. Water and juice bottles are not, and they may burst if you put compressed air into them.

Gather tools

- ▶ Hacksaw or PVC cutter
- ▶ Drill
- ▶ Drill bit the same size as the valve stem
- ▶ Circular file smaller than the valve stem
- ▶ Flat file
- ▶ Pliers
- ▶ Sandpaper, around 80 grit
- ▶ Safety glasses
- ▶ Bicycle pump, hand or foot powered
- ▶ Hot glue gun and hot glue

TINKER

Lay out the parts and cut PVC lengths more or less as shown in this photo. Naturally, you can change any part of it you want. It's more convenient to have a nice short one, but short barrels are less accurate, and the bullet doesn't have time to get going very fast either, so you don't want it too short.

We'll go first to the trickiest part. This is a common strategy when tinkering: if you leave the

most challenging part to the end, and then you find out that it's not just challenging but actually impossible given your current stuff and skills, then you just wasted a lot of time doing the other parts.

The trick in question here is to attach the soda pop bottle to the system as the reservoir of compressed air. This is not easy because its plastic is not the same as the plastic of the PVC.[1] Thus, you can't use the PVC cement. We'll use the 5-minute epoxy, but first we'll have to make the PVC fit nice and tightly in the mouth of the bottle. Usually the ½-inch PVC in the United States is slightly smaller in diameter than the mouth of a bottle, so what you need to do is create a bulge in the PVC so that it jams in tight.

You can make this bulge with a candle. You could also do it with a stove, but you'll want to heat up a small area around the circumference, so a pointy little candle flame is just the thing.

> **SAFETY NOTE** Get a friendly adult to help you with this part, and have the fire extinguisher handy. Also, don't burn the PVC, and do this in a well-ventilated area so that you don't breathe any fumes given off from the hot PVC. We had a fan blowing on us as we did this.

This part looks a lot more complicated than it is. You just hold the PVC above the flame—not *in* the flame—and turn it slowly. When the PVC is a little bit hot it gets soft, and then you gently push it together until it bulges a bit. It only takes a few seconds. Count to 12 and see if it's ready. If not, try another 10 seconds.

You don't have to push very hard; if you push too hard, it will buckle and fold. Try to keep it straighter than Wiltor did in that photo, and maybe hold it a bit higher over the candle than he did so that it doesn't get as black.

When you've got a nice little bulge, hold it until it cools and hardens again. You can blow on

it if you want. Then wipe the soot off a bit and jam it into the bottle mouth to be sure it fits tightly.

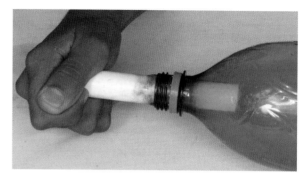

The next step is to take it back out and rough it up a bit with sandpaper. This will help the epoxy glue bond to the PVC. Hard plastics are some of the most difficult materials to get glue to bond to, and roughing them up often helps.

In Timor, for whatever cosmic reason, the main type of PVC available is gray in color and the size they call "½ inch" is slightly *larger* than the mouth of the average soda bottle. (I got ahold of a piece of the smaller white stuff just to show you how to do the bulge.) We make our compressed air guns from this larger, gray PVC, so I'll now show you what to do if you happen to end up with PVC that's too big.

Light your candle and heat up the center of a small piece, rotating just as if you were going to make a bulge.

After 12 seconds or so, when you think it's soft, start *pulling* on both sides, until it comes apart right at the hot spot.

Again, this is not difficult at all and takes only a few seconds. But you have to move quickly while it's still hot and soft. Take one of the pieces and jam the soft, hot end, which is now slightly smaller than the normal diameter of the PVC tube, into the bottle mouth. It should jam in nice and tight.

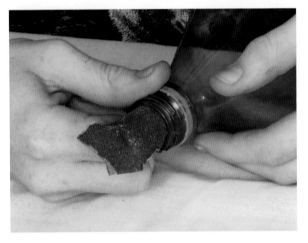

Mix up your 5-minute epoxy glue and slather some onto both surfaces, and then press them together tightly.

After a few seconds of cooling, take it out and rough it up. Rough up the inside of the bottle mouth as well.

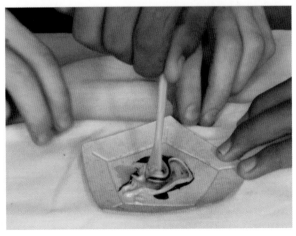

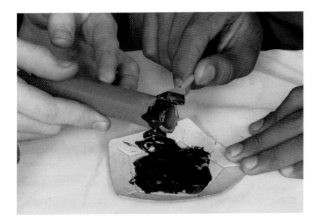

The other tricky part of this project is getting the valve stem into the PVC pipe cap. First cut off a valve stem and trim away the base until it's small but not gone. Rough it up with the sandpaper too, right around the base where the glue will contact it.

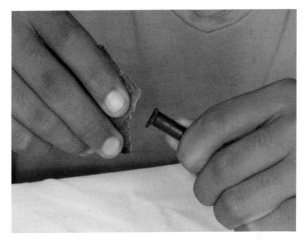

Now get a drill bit that's about the same size as the valve stem or slightly smaller and drill a hole into the pipe cap, right in the center.

Pack more of the epoxy glue around the joint and lay it aside to dry without bumping it. It says 5 minutes, but give it at least 10.

It's best if this is a super tight fit, so if you can't get the valve stem in, use a circular file to enlarge it bit by bit until the valve stem just barely squeezes its way into the hole.

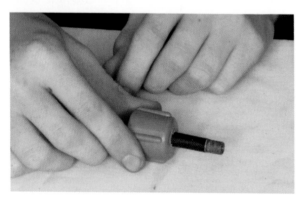

Now do the epoxy routine again, putting it on both parts and jamming them together, then leaving them to dry without disturbing them. Avoid getting any of the glue in the hole the air will go into.

Put the pieces of the gun together as they're going to be. Don't shove them together too tightly, because you're next going to take them apart for the PVC cement. But you do want to

check the whole arrangement before you do the cementing—it's a permanent bond.

If you're happy with it, go get the PVC cement![2] This is another source of nasty fumes, so go outside.

> **SAFETY NOTE** Use PVC cement in a well-ventilated area, and don't breathe the fumes.

Here again you'll want to put liberal amounts on both the inside and outside, then jam them together. Be sure you get the angles right on the T and the ball valve. Hold it up before cementing each joint so that you get it how you want it to be.

When it's done, check each joint to be sure you didn't forget any. Then put it aside for at least an hour—better yet, several hours. You don't want those joints coming apart when you pressurize it.

The last step on the gun is to extend the handle of the valve. This is not really necessary but helps to get a nice clean shot off. We use bamboo, as usual here in the tropics, but you could use a little stick of any kind. Put it up parallel to the handle and figure out which side you want long and how to strap it down with the hose clamps.

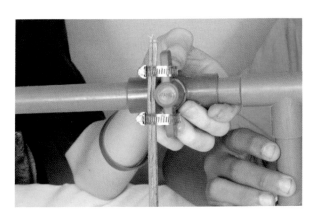

Later we'll put some tape on it to make it look and feel good.

Now you're going to need some projectiles. As mentioned, these guns traditionally shoot potato bits, but that's a no-go idea when you're living in a land where hunger visits all too frequently. Fortunately, in this land there is no shortage at all of old flip-flops.

To make bullets out of flip-flops, you need a short piece of PVC the same size as the gun barrel. It helps to sharpen the end of the piece so it can cut into the flip-flop better. You can sharpen it with the file, going around and around the outside of the tip.

Then twist it into an old flip-flop, or any other piece of thick, soft rubbery stuff, and pop a piece out. Once it's out, you may want to trim it up a bit more with scissors. It should fit tight but not too tight into the end of the barrel of your gun.

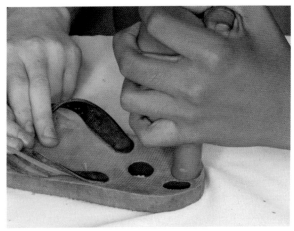

You can make a bigger projectile by hot-gluing two of these flip-flop bullets together. Again, trim the resulting double bullet with scissors so it is just the right tightness.

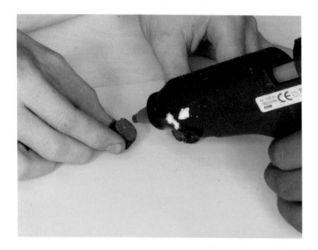

When the PVC cement has all hardened and you've made some bullets, you're ready for some

artillery action. Find a stick to use as a ramrod, and mark it so that you can push the bullet down near the valve without actually entering the valve.

Grab the bicycle pump and your safety glasses and go somewhere far away from any other human or animal life. Hook up the pump to the valve stem and pump it up to 20 psi, that is, pounds per square inch.

SAFETY NOTE Always wear safety glasses when using your compressed air gun. Never pump it up past 20 psi. Never point it or shoot it at anyone. Never put in any sort of hard or pointed bullet. Don't hold the bottle near your face or head. Be super careful with this thing—all fun ends when someone gets hurt.

Wait, there's more! If you grow tired of the random inaccuracy of fat cylindrical bullets, you can make rockets to launch with this as well!

Wrap paper around a piece of PVC to make the body of the rocket. Tape it off so that it's snug but not too tight on the tube.

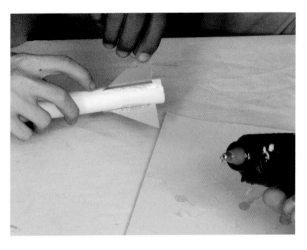

Now, this one is slightly more dangerous because it's pointy. It's best to shoot it straight up, and let it come down away from everyone. Keep the bottle away from your face.

Form a cone and fins for the rocket and tape or glue them on.

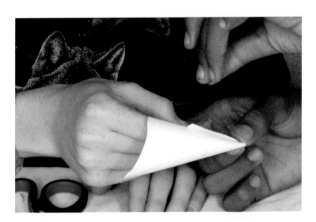

So, we've gone a bit nuts with this idea. Paulo and his buddies made this three-shot unit:

You can see how all three bottles have valves, so you open them and pump the whole thing up, then close them all. When you're ready to shoot, open them one by one, and you get three shots with the top valve. Or open them all and get a shot with three times the air pushing on it.

We also made this one with exchangeable cartridges, separate charger unit and no valve stem on the main gun body.

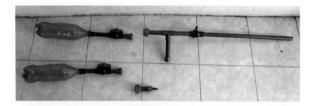

Here's one cartridge being pumped up. You then shut its valve and hook it up to the barrel. Step aside, James Bond.

Finally a pistol with a single-serve drink bottle. It doesn't pack much punch but handles quite nicely.

WHAT'S GOING ON?

Pressure is one those words people throw around without knowing the exact meaning in physics. Here it is: force per unit area. That means when you put 20 psi of pressure into your launcher, at every place on the inside of the bottle, as well as the inside of the tube and hose to the pump, and even the valve itself, you can mark a little square of one inch on a side and that square will be supporting 20 pounds of force.

To understand this better, let's think of something else that gives 20 pounds of force. Well, a 20-pound weight gives 20 pounds of force. Weight is the force that gravity gives. Let's think of water: 1 liter is 1 kilogram of mass, which weighs about 2 pounds here on Earth, so 10 liters would be 20 pounds of weight. So five 2-liter soda bottles would be around 20 pounds of force. Or if you want to think in terms of gallons (like gallon jugs of milk), there are around 4 liters in a gallon, so 20 pounds of force is about 2 and a half milk jugs of weight.

That's the force then. Pressure defines the area that receives that force. In our example, it's a square inch. So imagine 2 and a half milk jugs standing on a 1-inch square area, like maybe a postage stamp. That would be 20 psi. That's the force felt everywhere inside your air gun system.

That's significant. The thin plastic of the bottle can handle this, as can the PVC of the launcher tube, but you can see how you wouldn't want to increase it. At some pressure the system won't be able handle it and something will give way, and you'll then have an explosion with flying shards of plastic.[3]

Back to our shooter: Let's work out how much force the bullet gets. A ½-inch PVC tube has an inside diameter of about half an inch, by golly. The area of a circle is equal to pi times the square of the radius, which is half the diameter, here ¼ or 0.25 inches, so you've got

$$A = \pi r^2$$

$$A = 3.14 \times (0.25)^2$$

$$A = 3.14 \times 0.063 = 0.196 \approx 0.2$$

That wiggly equals sign means "approximately equal to," which is what we assume when we tinker. So the answer is 0.2 square inches, and that's the fraction of the 20 pounds of force that the bullet gets. We can take 0.2 times 20 psi and get 4—that is, 4 pounds of force pushing that bullet forward when you open the valve.

And how fast does it get going? Ah, that depends on how long the 4 pounds is pushing on it! As long as the force keeps pushing on the bullet, it will go faster and faster. (See Newton's second law in Chapter 19.) Now you can see how rifles get bullets moving faster than pistols do: the longer barrel gives the bullet a longer time to accelerate. The key point here is that when the bullet just emerges from the barrel and is ready to go flying through the air toward the target, it is going as fast as it will ever go. After that, it begins hitting air and eventually the target, or the ground, or something else that gets in the way, and all that stuff gives it forces that eventually bring it to a stop. (See Newton's first law.)

I did the calculations and got around 35 miles per hour maximum if it's accelerating for 1 foot of the barrel, and around 50 miles per hour if you had a 2-foot barrel. So it would just about keep up with the family car. But this is just a rough calculation, and the actual speed is almost certainly less, due to friction inside the barrel, leaks around the bullet, a drop in pressure as the bullet moves forward, and so forth.

There's a whole bunch more physics happening in your flip-flop air gun, but I'm running out of space here, so I'll just mention a few more things you could look into if you want. You can watch the shape of the projectile's trajectory and compare it to other things you've seen thrown or shot. You can shoot the flip-flop bullet straight up and measure the time it takes to return to Earth, and then calculate its maximum height with simple motion equations.

Or you could just go shoot cans off a bucket in your backyard all afternoon. That's tinkering too!

ENDNOTES

1. PVC stands for polyvinyl chloride, a group of organic molecules called polymers with the form $(C_2H_3Cl)_n$, where n could be any number. The soda bottle is very likely PET plastic, which stands for polyethylene terephthalate, a different group with the form $(C_{10}H_8O_4)_n$. Each of these is produced from petroleum and pervades our lives in the form of clothes, construction materials, and packaging among many, many other applications. PET happens to be easily recyclable, and PVC happens to sometimes deteriorate to dioxin, a heinously toxic substance, when burned. Various initiatives are in the works to reduce the use of PVC. But we'll just use a little bit to make our compressed air guns and Pipes of Pan before it is banned...

2. You may wonder why one of these is called glue and the other cement. Glue sticks two surfaces together by binding to each one. Cement actually alters the surfaces so that they bind directly to each other. Another place cement is usually used is in patching tire tubes, as you'll know if you ever patched your bike tire. You put rubber cement on the tube with the hole and sometimes also on the patch and you can watch the rubber get a bit slimy. Then you press the two surfaces together and the rubber pieces actually bond straight to one another. That's if you get a good bond. A bad bond will peel right off, because it's just stuck with the stickiness of the rubber cement.

3. Some young genius was doing a late-night dorm-room experiment in my university and his bottle blew up midway through. Fortunately it was near his leg and not his face (safety glasses were not part of his lab apparatus), and a piece of the shrapnel from the bottle lodged itself deep inside his calf muscle. I think his motto was "Safety second!"

Chapter 19

NEWTON'S SLINGSHOT

Tinker Newton's laws with rubber-band power.

Kids in Timor love to make and shoot sling-shots. As I watch them pull the rubber bands back and set the stone to sailing, I always think of Uncle Isaac Newton's laws of motion. Here's a fun little set of experiments we developed for teachers to demonstrate those laws in the context of slingshots.

Gather stuff

- Forked branch
- Rubber bands, maybe 20
- Piece of sturdy cloth
- Loop of yarn or a stick to scratch a circle on the ground
- Rocks: two the same, one big and one small

Gather tools

- Saw
- Scissors

SCIENCE IDOLS

Your Uncle Isaac Newton was one of the bright stars of the European Scientific Revolution. In the midst of the Renaissance, the Age of Discovery, he made phenomenal advances in the physics of force and motion—our topic here—as well as astrophysics, light, and color, and in his spare time developed the entire branch of mathematics called calculus.

That said, he was an incredible elitist as well as something of a social mad dog, arguing interminably with other scientists, jockeying for aristocratic positions, and relying on his family's wealth. Some of his findings were simultaneously discovered by others, with whom he fought until his death to claim sole credit, even though we have ample evidence that he used any and all data he could get from other scientists as well as from artisans and craftspeople—none of whom he gave credit. Ah, well, uncles can't be perfect.

It's best not to idolize these famous scientists. I say this because I've seen this sort of idolization happening all over the world. In schools in China, India, and Indonesia I've seen whole sets of posters, ostensibly documenting the greats of scientific discovery but that consisted entirely of European men. It's just not an accurate representation.

First off, as Uncle Newton himself said, scientists make progress by standing on the shoulders of giants. In many cases, those giants were from China, India, or the Islamic world. When Europe was having an embarrassing meltdown called the Dark Ages, Islamic scholars were developing algebra with a number system from India far superior to anything in Europe (ever tried long division with Roman numerals?). China made hundreds of scientific and technological advances *before* Europe, in some cases hundreds of years before.[1] Did your history or science teacher tell you that?

I love stories of the great scientists as much as anyone. I just don't give them all the credit. You see, you can't do science in a social vacuum. Your work always depends on the people around you and the society you live in. Like this activity for example: carry a slingshot and you'll be a target for other slingshots! In political science, it's called the Security Dilemma. Proceed carefully!

TINKER

Make a slingshot! Cut a nice sturdy forked branch from a healthy tree. Get a bunch of rubber bands and a piece of cloth.

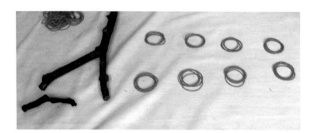

We used these skinny rubber bands in sets of four, hitching them together with that knot where you put one side of the loop through the other loop and back through itself. Use the same one to tie the rubber bands to the stick, preparing it on your fingers and then sliding it on the ends.

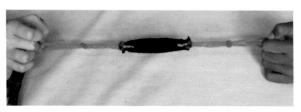

Uncle Newton wrote three laws of motion. We'll examine them all, but it so happens that the first one is the most subtle and the third one the most obvious, so we'll do them in reverse order.

NEWTON'S THIRD LAW

Put a rock in your slingshot and draw it back. Don't shoot. Feel the forces on your two arms.

What's Going On?

There you go: it's the third law:

For every action there is an equal and opposite reaction.

By *action*, Uncle Newton meant *force*. You want to give a rock a force forward, so it flies forward and hits something, right? The hand holding the slingshot is exerting a force forward, right?[2]

But now look at the back hand, the one holding the rock. Its force is directed backward. The forces are equal, as I'm sure you can appreciate; you're pulling forward with the same force as you're pulling back.

This, my friends, is always, always the case. Any force you find has another force opposing it:

▸ Gravity pulls a book down onto the table, and the table pushes up on the book.

▸ You push against a wall and the wall pushes back on you.

▸ The back wheel of your bike pushes back on the road as you pedal, and the road pushes forward on the bike.

▸ The propeller on an airplane or boat pushes the air or water back, and the air or water pushes the airplane or boat forward.

You can already see the problem when physics textbooks speak of "action-reaction forces," as if they're some special subset. It's all of them, buddy, and don't you forget it: forces always come in pairs. Now on to the second law.

NEWTON'S SECOND LAW

$$F = ma$$

Any questions? Usually the law is stated with mathematics like this, because everything we'll do in the following long-winded, highly articulated, ultra-analyzed tinkering is contained in that equation. Of course, for you to understand it, you have to know the mathematics called algebra. If you aren't fluent in algebra you won't fully get it yet. Not to worry; read on.

First off, the three letters represent three variables: F is force, m is mass, and a is acceleration. These are three quantities that can vary in a given motion situation, and they are oh-so intimately related. The m and the a are right next to each other, and that means they're multiplied in the equation, so you could also write $F = m \times a$ or $F = m*a$.

We'll do three pairs of slingshot shootings here, each time holding one of those variables constant and seeing what happens in the relationship between the other two.

#1. Launch the big rock and then the little rock, pulling the slingshot back to the same distance and keeping it level. We used 25 centimeters. You may want to practice launching the big rock a few times; it's not as easy as the small rock.

We saw the small rock go much farther than the big rock.

#2. Get out the two rocks of the same size. Shoot the first one by pulling back on the slingshot just a bit, and the second one by pulling back a lot. Keep the angle of the slingshot as steady as you can.

I'll bet you saw the first rock go a short distance and the second one go farther.

#3. Now make a circle on the ground, either with a loop of yarn or rope, or just scratching it in the dirt. (Ours is a bit skinny—you can barely see it beside the pumpkin vine.) Shoot the big and small rocks until you hit the circle. As you do this, note how far you had to pull the slingshot back for each of the rocks.

You don't see it in the photos, but Zaya had to pull the big one back farther to get it into the circle.

(I hope you don't get tired of me telling you this, but please *don't trust us on this*. I mean really; you don't even know us! We could be lying straight-faced to you. What's more, this is so easy to try. So do it and place your confidence squarely in yourself, not in these two girls of questionable motive...)

What's Going On?

Uncle Newton's second law lays out the relation between three variables. Force is a push or a pull, mass is a lot like weight—that is, how heavy something is—and acceleration is how fast the speed of something is changing.

Acceleration is the complicated one. Speed is easier—that's how fast something is going. Acceleration is the change in speed. So when you put your foot down on the accelerator, you start to go faster and faster, and you're accelerating. And when you hit the brake, you start to go slower and slower, so you're accelerating again, this time with a *negative* acceleration.

But when you're doing 65 miles per hour down the freeway, needle steady on the speedometer, radio blasting, breeze blowing back your hair, not a care in the world, your acceleration

is...zero. And then, when you get pulled over by a cop because it was a 50-mile-per-hour zone, once you come to a stop your acceleration is again zero. Anytime your speed is constant, no matter what the value is, your acceleration is zero.

Now let's look at the relations we just saw one by one. In #1, the force was constant. The rubber bands give force depending on how far you pull them back, so if you pulled the slingshot back the same distance both times, it gave a constant force to the big and the small rocks. And if your rocks sailed like ours did, the small one went much farther.

How far the rock went is sort of a measure of acceleration. It's really hard to measure otherwise, but all else being equal, the rock that had a higher acceleration in the slingshot is going to go farther. So this is an easy way to compare accelerations. In physics we say the distance is *proportional* to the acceleration. So this means the big rock had a smaller acceleration than the small rock.

So there is the first relation:

1. When force is constant, the bigger the mass, the smaller the acceleration.

In #2, the mass was constant; the two rocks weighed almost the same. You gave one a big force by pulling the rubber band back a lot and the other one a small force by pulling the rubber band back a little. If yours sailed like ours did, the one getting the larger force went far and the one getting the small force didn't go as far.

That's our second relation:

2. When mass is constant, the bigger the force, the bigger the acceleration.

Finally, in #3 acceleration was constant. You shot them to land at the same distance, so their acceleration had to be more or less the same. If your results were like ours, it took a larger force to give that acceleration to the bigger rock—that is, you had to pull the slingshot back farther. You had to pull the slingshot back only a short distance to give the smaller rock that acceleration.

The final relation then is this:

3. When acceleration is constant, the bigger the mass, the bigger the force.

Again, all that verbiage is contained in the elegant little equation F=ma. Once you understand algebra like this and even higher math, you can recognize one of the great wonders of the universe that many scientists have marveled at: how does this language called mathematics fit so well with our observations of natural phenomena? I'll leave you marveling on that and wrap up this second law.

All those relations I wrote can be written the other way round as well. You could call it the vice versa form:

1. With a constant force, the smaller the mass, the bigger the acceleration.

2. With a constant mass, the smaller the force, the smaller the acceleration.

3. With a constant acceleration, the smaller the mass, the smaller the force.

You'll notice that in the second two relations, the variables go together: the bigger, the bigger; the smaller, the smaller. This is called direct proportion. Only the first one has variables in what we call an inverse proportion: the bigger, the smaller; the smaller, the bigger.

This all makes perfect sense, and I think if you had made predictions of the results of this experiment, you would have nailed most of them. Here are some examples from daily life that follow those rules and just make sense if you think about them.

- A big truck (big mass) is going to require a bigger engine (bigger force) to reach

a certain (constant) acceleration than a motorcycle (small mass).

- If two identical cars (constant mass) start off together, the one that wants to reach higher acceleration in a certain distance will need a greater force.

- If you give the same push (constant force) to two kids on tricycles, one big husky kid and one little skinny kid, the skinny one will take off faster (higher acceleration).

NEWTON'S FIRST LAW

Okay, ready? Here it goes: grab the slingshot and shoot a rock, any rock. (Don't hit anything delicate!) Watch what happens.

What's Going On?

Did it come to a stop in the end? Of course it did! That just makes sense right? If it hadn't stopped you'd have freaked.

Around 2,360 years ago, Aristotle wrote down what we all know full well: objects in motion tend to come to rest. Even more obvious than that, he wrote that objects at rest tend to stay at rest. End of story, right?

Well, actually, no. Some 1,700 years later Uncle Newton made several interesting observations, along the following lines:

- A slingshot rock does *not* come to rest until it hits something. As long as it doesn't hit anything, it keeps on flying.

- The moon doesn't seem to have a rocket powering it to orbit around the earth, nor does the earth seem to have a rocket powering it around the sun. It seems celestial bodies just tend to move forever without any added force.

- If you roll one marble such that it smashes into a row of marbles, it will stop and the marble on the end will take off rolling with about the same speed. It will often keep rolling until it hits something.

These and various other observations led Uncle Newton to formulate his law quite a bit differently than Aristotle's. Here it is:

Objects in motion tend to stay in motion, until they receive a force to change that motion or bring them to rest. Objects at rest tend to stay at rest, until they receive a force that sets them in motion.

Today you can see videos on the Internet in which astronauts in orbiting laboratories pass things to each other with the slightest flick of a finger. The object heads slowly across the spacecraft until it arrives at the other side. Had Uncle Newton seen such footage, he may have come up with his law even earlier (he probably didn't have Wi-Fi). It's all obvious when you escape the earth's gravity.

So that force pulling everything down messes with our observations. Even when something round is rolling on the ground so it's got no sliding friction, it hits little bumps and gets friction from the air, and in the end always encounters enough forces to bring it to a stop. We hardly get to see any examples of an object in motion staying in motion. We can forgive old Aristotle: he had so few examples to observe.

That was the brilliance of Uncle Newton. He saw that the moon and planets were the norm—moving on and on without stopping—and that everything here on Earth was getting a stopping force when gravity brought it in contact with Earth.

You can see this law in action in situations of super low friction, like the rolling marbles, or an air hockey table, or something heavy sliding on ice, but the best example of objects staying in motion are projectiles. The farther you shoot them, the farther they go, and if you shoot one far enough, the earth's surface will fall away due to its curvature and the object will end up falling *around* the earth indefinitely because it never gets the stopping force. That is the situation of a satellite.

And those are Uncle Newton's three laws of motion! Next time you throw, hit, or kick something into motion, think about how the forces relate to these laws. It doesn't matter what hemisphere you're in, or even what planet or galaxy; they apply everywhere!

ENDNOTES

1. Check out Joseph Needham's seven-volume *Science and Civilization in China*, or *The Genius of China*, a one-volume overview by Robert Temple.

2. We're not talking motion on this third law, just force. But we could talk motion as well. When your hand releases the rock, the force the rubber bands give to the rock will send it sailing forward, and that front hand will continue to give its forward force, steadily diminishing, until the rock has left the rubber bands. Then it will no longer have to give any force at all. More on this in the second law.

Chapter 20

FLOATING AND SINKING

Play with stuff in water to get a feel for Archimedes's law.

Here's a classic physics question we ask high school teachers in Timor. It involves a big rock in a boat that is floating on a small pond. A post stuck in the bottom of the pond extends up above the surface of the water, and the level of the pond can be seen on the post. The pilot of the boat picks up the rock and chucks it out of the boat into the pond. After the waves calm down, the pilot peeks at the post to check the level of water in the pond. What does she find?

- The water line went down, meaning the pond level decreased.
- The water line went up, meaning the pond level increased.
- The water line stayed the same.

And the answer is...C'mon, I'm not going to tell you the answer when you can do the experiment so easily with a couple of bottles and a rock!

Gather stuff

- Big bottle, top chopped off
- Small bottle, top chopped off
- Water, to fill the big bottle halfway

- A decent-sized rock that fits in the small bottle and doesn't sink it

Gather tools

- Scissors

TINKER

Fill the big bottle half full. This represents the pond full of water.

Float the small bottle on the water in the big bottle, and then gently drop the rock into the small bottle. This represents the rock in the boat.

Mark the water level on the outside of the big bottle. This is the water level referred to in the question.

Also mark the water level on the outside of the small bottle. We noted this while it was floating but then had to take it out before marking it. This does not represent the water level in the question, but rather the water level on the side of the boat. That's interesting, too, so it's good to watch what happens to it as well.

Now take the small bottle out of the big bottle, careful not to drip any water outside the bounds of our ultra-high-precision scientific experiment; remove the rock; and drop it gently into the water in the big bottle—no splash.

Drop the small bottle back into the water.

Check for change in water levels on the big and small bottles.

Any questions? The result is pretty obvious, eh?

Wait, you haven't done it yet? Well, get up off your chair and go do it! You know the materials are waiting in your recycle container!

WHAT'S GOING ON?

We do this activity as we're dealing with Archimedes's law, which tends to come up in textbooks:

> The buoyant force on a body in a fluid equals the weight of the displaced fluid.[1]

You may recall the great (and highly unlikely) story in which Archimedes was sitting in his bath, stark naked, circa 200 BC, pondering a challenge the king had given him: to prove his recently fabricated gold crown was not tainted with other metals. Watching the water he displaced rise on the side of the bathtub was the breakthrough moment for the venerable old thinker, and he leapt from the water and went running down the street, stark naked, shrieking "Eureka!"

What he'd realized in the fanciful story and what he wrote about in a treatise (which we do have pretty good evidence was his) was that you can calculate the volume of an irregular object by submerging it in water; the rise in water will be equal to the volume of the object. In our experiment then, when you drop the rock directly into the water, the water rises by a level equal to the volume of the rock.

But what about when it's in the boat? In the boat, the rock can displace *more* water. Check this out: Anything floating in water is getting an upward force—buoyant force—equal to the force of gravity on it. Because these forces are balanced, it floats. If the force of gravity is greater than the buoyant force, the thing sinks. If the buoyant force is greater than gravity, the thing shoots up in the water.

For a floating object, these up and down forces are equal because the thing is displacing a weight of water equal to its own weight. Some part of the floating thing may be sticking up above the surface of water, but the *part* that is *below* water level has a volume equal to the volume of water it is displacing, which weighs the same as the entire object. Better reread that; it's quite a concept.

Take an empty boat floating on the water:

The part of the boat under the surface is displacing a certain volume of water, and that volume of water weighs the same as the boat. There is some air and maybe some ropes and nets and suchlike inside the boat, all under the water level, but the material doesn't matter here; it's just the total volume of water displaced that matters.

Now toss in a pile of rocks:

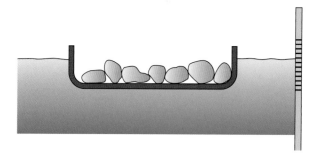

The part of the boat under the surface has increased so that the displaced water has increased, which makes sense because now that volume of displaced water has to weigh the same as the whole boat and the additional rocks.

Now pile in more rocks until the water is right up to the rim of the boat (called the *gunwale* in boating terms):

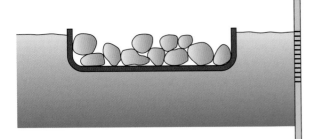

The boat has sunk down deep into the water and is now displacing its maximum. Still, the water displaced is of a weight equal to the boat plus the rocks.

Now toss in a few more rocks and watch in horror as the water flows over the gunwale into the boat and it sinks to the bottom of the sea.

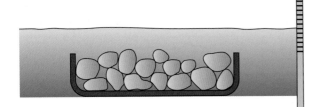

What happened? The boat had no more means to displace more water, and the weight of the water displaced was less than the weight of the boat plus all those rocks. Archimedes's buoyant force depends on the weight of that displaced water, and since it's now less than the weight of the boat and the rocks, gravity wins and pulls it under. (It does not depend on whether or not you're naked.)

You can see that a bigger boat would have the capacity to displace more water and carry more rocks. The main point for our experiment is that being in a boat allows a rock to displace

more water—in fact, to float—whereas without a boat the rock just sinks.

Meanwhile, the water rises in the pond according to how much water was displaced. So when it's just the rocks, they displace less than when they are in the boat floating up at the surface. The water level on the post went down after the boat pilot chucked the rock into the drink.

To summarize, floating and sinking depend on two forces: *buoyant force* going up and *gravity* going down. Gravity depends on an object's mass (and the earth's mass, but we're somewhat limited in our ability to change the earth's mass or move to another planet for the experiment). Buoyancy depends on the displaced water, like old Archimedes said, which is determined by the object's volume.

Another way to think about this is to imagine (or make...) two identical sealed bottles, one filled with sand and one filled with air. Under the water, they'll displace the same, because they're the same volume, and so will have the *same buoyancy*. But the one full of sand is much heavier, heavier than the buoyant force on it, so gravity wins and it sinks. The other is so light that it needs only a trifle of the total buoyant force possible, so it sinks into the water only a trifle and floats merrily on the waves.

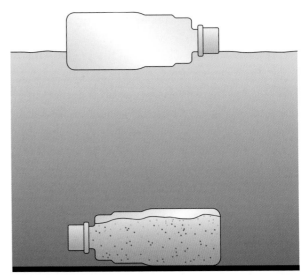

Now think of a small rock the same mass as that empty bottle. You could find one easily if you had a two-pan balance, or you could use a spring scale or any little scale. The bottle is much bigger than the little rock—that is, with a bigger volume, though the masses are the same. Chuck them both in the ocean, and you know which one will sink![2]

So what determines if something floats or sinks? Well, it's not the mass, and it's not the volume; it's both! That combination of mass and volume is so important it's been given a name: density. When we say iron is heavy and Styrofoam is light, what we really mean is that iron is dense and Styrofoam is not. Of course a giant car-sized chunk of Styrofoam is going to be heavier than a little iron nail, but the iron is still the denser of the two. That is what makes it sink, even though it's so small and light.

This is a concept that has serious consequences for Timorese fishers. Many small outrigger dugouts used in Timor don't rise much at all from the water. If the fisher catches enough fish to fill the boat, or if a leak is sprung and the incoming water not bailed out in time, the boat will soon fill with water, and the boat will be displacing only the thickness of the tropical hardwood itself, at which point it is not so much a boat as a piece of driftwood with two outriggers and a worried pilot.

Balloons filled with helium and hot air also float, in air. Just like boats, the buoyant force on balloons depends on their total volume, and the net force they feel depends on the difference between their density and the density of the air they're floating in.

Specifically, the buoyant force is a result of the difference in pressure above and below an object. Thanks to gravity, pressure decreases as you move up from the bottom of the ocean all the way to the top of the atmosphere. This pressure differential gives rise to buoyancy. Crazy, eh? Gravity pulls down and also makes fluids push lower density stuff up. If you ever make a boat, you'd better find a way to get its density lower than the water you hope to float on!

ENDNOTES

1. In case you were wondering, that's killer hard to say in Tetun, due to the lack of a word for "displaced." Even the word *buoyant* is no problem to translate into Tetun: we just call it "forsa ba leten," meaning "upward force," and it's usually obvious that we're talking about buoyancy and not lift. Lift happens in airplanes and other things moving through fluids. Buoyancy happens on anything sitting in a fluid.

2. Say, if you did actually do that experiment, do us all a favor and take the blooming bottles out of the water when you're done. At last count, 15 million tons of plastic per year is flowing into the ocean and destroying the ecosystems there, whether sinking or floating.

Chapter 21

LAKADOU

Make a sonorous knock-off of Timor-Leste's amazing bamboo stringed instrument.

CREDIT: Ros Dunlop

Timor-Leste is home to several interesting musical instruments, among them the *lakadou*, which would fit somewhere between the string section and the percussion section in an orchestra. The cords stretched from end to end of the huge bamboo section are actually the skin of the bamboo itself, carefully cut and lifted out, and held to vibrate above the main tube by small bits of sticks at either end. The incredible part to me is how it is tunable by pressing on the stick bits. You strum it with another chip of bamboo or coconut shell, and it's a trick to strum in a way that hits many of the stretched strips. Then you get another person to tap out a rhythm on one end with two small sticks.

The strips cover maybe two-thirds of the circumference of the bamboo tube. At one end of the *lakadou* the section division of the bamboo seals the air inside, and the other end is open to be controlled by your hand. There's a hole on the underside halfway down that you can open and close against your belly to give subtle changes to the tones produced.

Such hefty bamboo is not easily found, even here, but a while back we realized we can make a fake version of the *lakadou* with string and a cardboard tube. We were surprised at how nice the sound is. Check it out.

Gather stuff

▶ Fat cardboard tube, maybe 40 or 50 cm long

▶ Strong, thin string

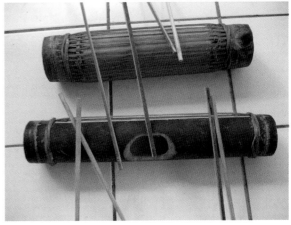

CREDIT: Ros Dunlop

- ▸ Small sticks, like chopsticks, ¼-inch dowels, or pencils
- ▸ Pick, store-bought or cut from an old credit card

Gather tools

- ▸ Hacksaw or cross-cut wood saw
- ▸ Scissors
- ▸ Side cutters

TINKER

Start by cutting the tube around 40 cm long with a hacksaw or small-toothed wood saw. Longer tubes will make deeper pitched sounds.

With the hacksaw, cut 1-cm slits on either end, directly aligned with each other. We started with five on either end. You can always add more later.

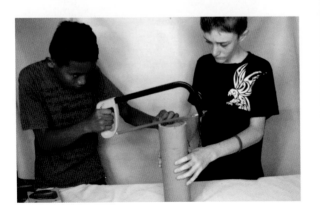

Cut a piece of string maybe twice as long as the tube and make a knot at one end. Any big, fat knot will work. Pull it tight and slide the string into one of the slits so that the knot gets stuck on the inside of the tube. Here I'll show the process for the second string; they'll all be the same.

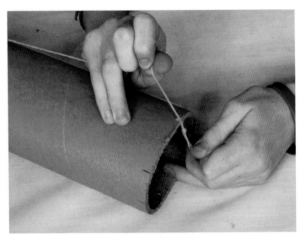

At the other end, pull the string to the corresponding slit. Form the knot and move it to a position just short of the slit.

Now pull like crazy so that the knot goes over the edge and the string falls down into the slot. The knot gets stuck on the inside.

Next, cut out your small stick bits. We sacrificed a pencil, which works well because it has flat edges due to the hexagonal shape. You could cut it with side cutters just as well.

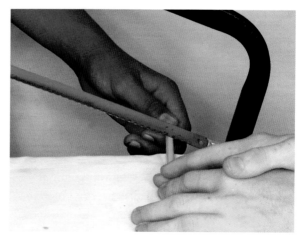

Jam one bit under each end of each string and cut out your pick, and you're ready to go.

Mount strings for all your slits.

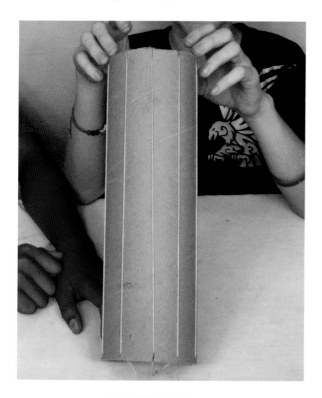

We didn't bother hacking a hole in the sidewall or covering one end, but you could. When you play it, you can cover one end with your belly and then open and close the other end with your hand to see what different sounds you can make.

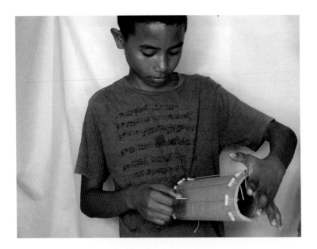

Jamming the sticks in or letting them out changes the pitch of each string. Sometimes the stick bits slip down. We solved that problem by rubbing a hot-glue stick where the stick sits on the tube to increase friction. This kept ours in place.

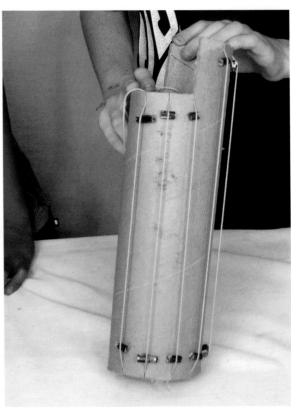

We made another one with two different lengths of string on a notched tube. Zowie!

As fantastic as we sound, the Timor talent scouts haven't found us yet.

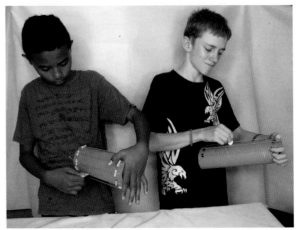

WHAT'S GOING ON?

Sound comes from something vibrating. In the case of musical instruments, usually the whole thing is eventually vibrating, but there is always an original vibrator. In this case it's the strings, representing the bamboo skin strips. They are stretched tightly, and when you pluck one, it goes back to its original position, keeps going due to momentum until it tightens up on the other side, and then heads back to the original position. But then it can't stop again and goes back to where it was released from the pluck, and then it starts the whole thing over again. It vibrates, in other words.

If it vibrates fast, it makes a high-pitched sound, and vibrating slowly makes a low-pitched sound. Three easy tweaks increase the speed of the string's vibration: stretch it tighter, make it lighter, and cut it shorter. Tweak vice versa to make the vibration slower: let it looser, make it heavier, and cut it longer. Try all of these on your *lakadou*, and also on your guitar if you have one. To make a string heavier, you usually have to get another one, maybe a used steel guitar string.

High and low pitch is about a sound's frequency. Loud and soft is about a sound's amplitude. If you hammer hard on something, it will vibrate more violently and make a louder sound. Play gentle and you have a soft sound. And there's some science of the music coming from your *lakadou!*

THE SASANDO

In West Timor and the nearby island of Roti, there is yet another instrument based on vibrations around a bamboo section. The *sasando* has steel strings like a guitar, tied at either end like our knock-off *lakadou*. It's got palm leaves cupped elegantly behind it that fold and unfold like a hand fan. I saw these when passing through West Timor on my way to Timor-Leste once; in Timor-Leste they are not to be found. Except by way of you-tube.com, where you can view jaw-dropping *sasando* clips of everything from traditional tunes to... "Despacito"!

CREDIT: Photo by Gunawan Kartapranata

Chapter 22

TUBE MUSIC

Make complex and curious sounds come from bamboo or PVC. CREDIT: Ros Dunlop

It's said you can hear the sea if you listen to a conch shell, even if you're not near the sea. It's not true—you have to be near the sea. You see, the conch shell receives the sounds around it and then amplifies some of them as they resonate inside its gorgeous, spiraling cavity. If you're near an interstate, you'll hear the interstate noise selectively amplified in there; it will sound a lot like the sea. If you're in a silent room, you'll hear nothing. Please don't take my word for it—try it yourself. You don't need a conch shell.[1] You can use a bottle, cup, or mug if you close off the mouth a bit, or you can even use your cupped hands.

It's the middle of the night as I'm writing this (as it tends to be for most of my scribbling), and I just checked this simple little experiment again. I was not expecting much because we're in a sleepy little cul-de-sac on an untraveled back alley and all the dogs have thankfully gone to sleep for the night.

To my surprise, I heard a pretty significant rush of sound, a lot like the sea, but steady! I looked around and it hit me: the fan! The fan produces steady white noise and the coffee cup I was listening to amplified some selected frequencies of that noise. A smaller cup selected different frequencies. Very nice.

All the instruments you'll tinker with in this chapter run on that same principle, with the addition that they actually *make* sound they select from when you drive wind past them with your breath. It's a pretty incredible process, being invisible and all, and can be quite pleasing to the ear. Have a go at it.

Gather stuff

- PVC, ½ inch, 2 meters
- Bamboo, if possible
- Assorted bottles, plastic and glass
- Pennies
- Aluminum can
- Straw
- Tongue depressors or similar flat stick, 2
- Tape—electrical and packing tape or duct tape
- Water
- Dowel, ¼-inch diameter, or similar stick
- Plastic bag, small and thin

Gather tools

- PVC cutter or hacksaw
- Hot-glue gun and glue sticks
- Knife
- Optional: drill and ¼-inch bit

TINKER

Bamboo is easy to find here, but PVC may be easier to find for you. Bamboo is also harder to work with; you can easily crush the sidewalls of the thinner stuff. But it does have inner walls separating the sections, so if you cut it just right, you don't have to plug the bottom of a section.

Tube Music Basics

Right off the bat, see if you can make a clear tone by blowing on a piece of PVC or bamboo.

Cut a tube around 20 cm and clean off the edges so there are no burrs to mess with the airflow.

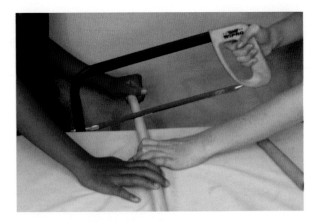

Make sure one end is plugged. You can just use your thumb, or you can tape a penny onto the bottom. (Shown here are Timorese 5 cent pieces and U.S. pennies, since it's an international sort of book.) Make sure the coin is taped tightly, with no wiggle room and no cracks for the air to leak out. We use packing tape because it's easier to find here than duct tape.

Now blow across the top opening. Rest your lower lip on the front of the rim, and blow smooth and steady and medium strong across the top in the direction of the far side of the rim. Change your angle up and down, and the shape of your mouth, until you get a nice, pure tone. If you are getting a breathy, vague sound, keep adjusting the wind you're sending out until you get the crisp tone.

If you can't get that nice tone, first make sure the tube is covered tightly at the bottom, and then try the following:

- Get a longer or shorter tube, or a ¾-inch piece of PVC.
- Try it with a bottle first; sometimes it's easier to get a good tone.
- Find some friends who can do it and mimic them.

Once you've got it, try the same thing with a tube that's not covered on the bottom.

You should hear a vague, breathy tone, but nothing crisp. Hmmmm.

While you're thinking about what's going on here, make a few more instruments. Pipes of Pan are just a set of closed-end tubes of varying length. Your Aunt Wikipedia has fascinating info to share with you on the pan flute. The Greeks really did use this instrument, whether or not the god Pan played it. The Andean cultures of South America call it the *zampoña* and have taken it to a high level of sophistication in their music.

Chop five or more sections of PVC or bamboo and cover the bottoms well.

Hook them together however you want. We use hot glue, which doesn't bond to hard plastic that well, so we back it up with tape.

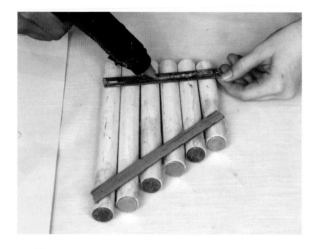

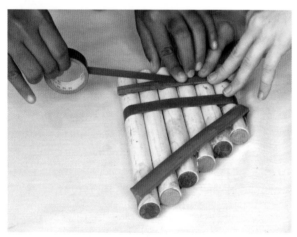

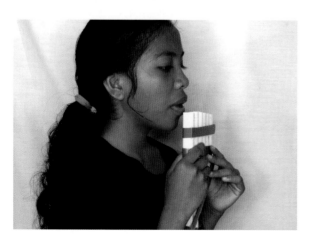

When you try playing them, you may notice that you don't have a nice scale but rather a random set of notes from low pitch to high. If you spent some time, you could tune your Pipes of Pan to produce a scale. Just choose the low tone you like, and then one by one, cut each consecutive tube a bit shorter, maybe 5 millimeters at a time, until you get the next tone of the scale. You could even use a guitar tuner and put it on a perfect pitch. You can download apps for this as well.

Before you move on, try a few things. Get a fatter PVC piece, like ¾ inch or 1 inch, cover one end, and blow on it. What do you notice?

Try covering the bottom of a tube with just tape—no coin. As you blow, put your finger on the bottom and feel what's happening. Press your finger tightly over the tape and then let it loose again. Pretty cool, eh?

One more thing to try is stopping up the bottom of the tube with water. Fill a wide-mouth bottle or jar with water and sink a PVC tube in it. Blow on it as you move it up and down.

Also try blowing a hard blast and listening carefully to the tone as you do so. All the while, you're trying to work out what the science is here—specifically, what determines the note coming out of the PVC.

The Slide Flute and Music from Other Vessels

Next we'll make a slide flute, which can give various tones. Chop and clean up a tube of about 40 cm. Get a stick that's half again as long as the tube and fasten the end of the plastic bag to the tip of the stick. You can use tape or tie it on, or slit the stick and slide the bag in a bit.

Wrap the plastic bag medium-tightly around the stick right near the end until you make a bulge that will just fit into the tube.

Trim off the rest of the bag. Then tape both ends down tightly, with some of the tape sticking to the stick. This is so the plastic doesn't slide up and down the stick as you push and pull on it.

If the top is very pointy, you can clip off a bit of the stick. You can tape over the bottom part of the stick too, for a nice handle. Now slide it into the PVC and try blowing on it.

Often it won't work, or it won't sound clear. Solution: add water! Using just a few drops usually does the trick.

Now you should get a clear tone that changes as you move the stick up and down.

Aside from the PVC tubes, you can also make this kind of instrument with bottles and cans.

Add water and the tone will change!

A can will make a sound if you use a straw to blow over its hole. You can take off the opener.

A glass bottle also makes at nice tone if you hit it with something, like a spoon. This tone also changes as you add water. Try it, and compare it to the changing tone you get when you blow. It's easier to dump water than to add it, so we start with the bottle three-fourths full, then tap and blow...

...dump some water...

...tap and blow again...

...dump some water, tap and blow again, dump some water, and on and on.

You hear the patterns? Are they conflicting in your ear, or your mind, or your mind's ear? More on this in the "What's Going On" section. Of course if you had eight bottles you could tap and/or blow a full scale and impress the socks off any audience.

A Real Live Flute

The final grand project is to make a PVC instrument like a concert flute, where one end is closed and you blow over a hole in the side wall near the end. Start with a piece of PVC around 30 cm long and drill one hole near one end, maybe 2 cm away, and several more farther down where your fingers can cover them up. Make the holes nice and clean.

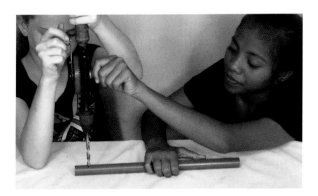

TUBE MUSIC MOUTHPIECES

Concert flutes have a bare a hole in the mouthpiece, and you're on your own to get your blow angle right so that it sets up a nice tone. But many tube instruments, like whistles and recorders, have a little structure, creating a path that directs the air across the hole. This limits the variety of sounds you can generate while making it a whole lot easier to blow a nice note.

Here in Timor, I've seen several ingenious mouthpieces involving bits of bamboo pressed to the main tube's outer wall. Here is one from Matata, Ermera.

Now cap the tube off. If you have a pipe cap, this is the time to use it, to attain the elegance necessary for future orchestral performances. If not, the coin and tape also work.

Now cover all the holes—damp fingers work well—and try to get a tone. Again, you have to move your mouth around, adjusting the angle and the shape of your mouth and the flow rate.

Don't be put off by how tricky this is; the flute is known to be one of the hardest instruments for beginners in a band. If you start getting dizzy, that's science too; your hard blowing and breathing is giving you more oxygen than you need, and your brain is going all giddy on you! When you get a tone, begin uncovering the holes and listen to the changes.

WHAT'S GOING ON?

Sound is created when something vibrates. The easiest example is your throat; say "Ahhhhhh" as you hold your throat and you'll feel the vibration in there. Guitar strings vibrate when you pluck them and piano strings vibrate when the hammer hits them. Clarinets and oboes have reeds that vibrate, and the whole brass section employs the musicians' vibrating lips. So what in tarnation is vibrating on a flute?

You're not going to be happy with my answer, because it's difficult to prove and almost impossible to see. It's the air itself.

One way to imagine this vibration is that when your breath hits that far rim of the tube,

some of it goes into the tube,

which increases the pressure in the tube,

and the increase in pressure tends to push air back up out of the tube,

and when the air comes flowing out of the tube it gets blown along with the part of your breath that's overshooting the rim,

but then when air comes out of the tube it decreases the pressure in the tube,

and the decrease in pressure tends to pull air into the tube,

and your breath is still there blowing so some of it goes into the tube,

and the cycle starts all over again.

I'm out of breath just having said all that. If your tube is resonating at around middle C, that's 261 cycles per second—that is, the pressure increases and decreases more than 200 times each second. Lickety-split.

If you get that, you can understand two more things. If the bottom of the tube is not covered, the pressure is never going to get very high because the air just goes out the bottom. You'll still hear a breathy note as the sound echoes around in there, like the conch shell/coffee cup note, but it won't build into a crisp note like the closed tube.

Also, you can see that if the tube/bottle/can is larger, it will take more air to build up the pressure inside, which will take more time, and so you can't get the cycles to happen as fast. This results in fewer cycles per second, otherwise known as a lower frequency, which gives a lower or deeper tone.

It's not at all easy to see this vibration, but feeling it is no problem at all: on the walls of a bottle when you're blowing a clear tone, and on the tape on the tube without the coin. You can see that the hard coins are useful at resisting the pressure change; the thin tape tends to move back and forth as the pressure rises and falls, which impacts the clarity of the tone produced.

Now think of all the ways you used to change the tone in this tinkering: different tube lengths, different tube diameters, a sliding stopper in the tube, holes in the side of the tube, water in the base of the tube, different-sized bottles, water filling up some of the space in the bottle, and so forth. In each case, you are tampering with the amount of air inside and thus the time it takes to turn around a cycle of vibration. I don't want to trivialize the entire brass and wind instrument sections of the orchestra, but that's more or less what the intricate structure of each complex instrument does as well. And what a glorious set of sounds results!

Did you get the point of the holes? When the air comes to a hole, it's just like it's gone out the top of the tube. So by covering and uncovering holes, you change the effective length of the tube and thus the vibration cycle.

How about the water in the slide whistle? That just helped seal the plastic plug. Any little hole will screw up the vibration because the pressure will not build up right. Try putting a tiny pin hole in the bottom of a plastic bottle. You'll lose the nice sound.

All right, then what about the bottle that taps a different tone than it blows? Careful—when you whack a bottle, the glass of the bottle is what's vibrating, not the air inside. If it is full of water, the water acts as a brake on the vibration of the glass, so the vibrations are slower and the tone is deeper. When it's empty, the glass walls can vibrate faster, so the tone will be higher.

Meanwhile, the tone you get from blowing air goes exactly the opposite: the less water, the more air, so the slower the vibration and the deeper the tone. Vice versa is also true: more water means less air, which means faster vibration, which means a higher tone.

Finally, I bet you already figured out the funny dwooping tone when you put the tube into the water. Now the bottom of the column of air is the top surface of the water inside the tube. When the pressure increased in the tube, it pushed the water down a bit, letting more air in and so dropping the tone. But the water didn't stay down—it surged back up, pushing out some air and so raising the tone again. Dwoop, dwoop, dwooooop.

ENDNOTES

1. And in fact, the conch is endangered in various places around the world, so think twice before you buy one of their shells, exquisitely beautiful as they are. I've seen the ugly reality here in Timor that a good number of the sea shells for sale were not found sitting abandoned on a beach somewhere, but instead wrenched from the backs of their soon-dead occupants. It's good money for the collectors, but bad business for the sea creatures, and another chip away at our fragile ecosystem.

Part VI
Electrons in Motion

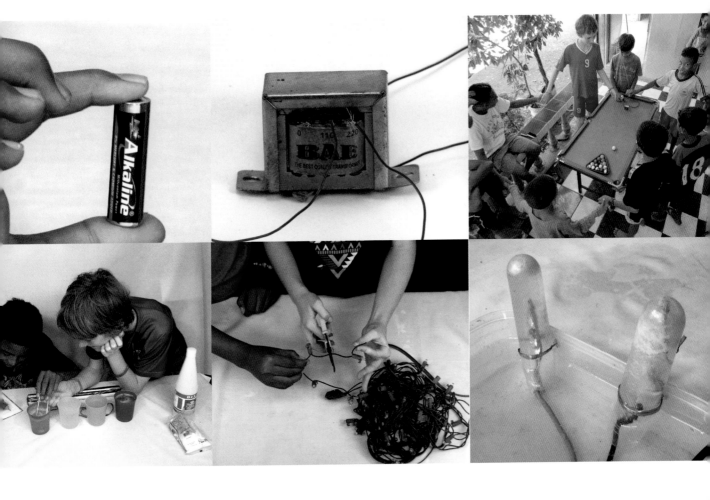

Chapter 23

ELECTRIFYING INDUCTION

Deliver a small shock with a common transformer.

Before we dive in to tinker, let's consider safety first. Electricity is invisible, like air, and quite a bit more dangerous. To live long into your golden years, you'd do well not to touch any random wires or electrical boxes, or anything electrical at all that's wet. Many houses here in Timor use small electrical pumps to draw water from shallow wells, and the combination of low-quality electrical connections, leaky water works, and no safety code has resulted in a heartbreaking number of deaths. So here is your warning:

> **SAFETY NOTE** You never, ever want to fuss around with a transformer that is hooked to the wall receptacle or that has any significant circuitry connected to it. You can feel safe with our activity because you're not hooked up to the wall electricity, so the only energy source is the small battery.

Now that we know enough to stay alive and healthy, let's admit that danger is exciting. Who's

not been thrilled by an electrostatic spark emanating from one's finger after scooting around the carpet in socks? My student in California came back from a visit to Mexico reporting a man in a local tourist park offering to shock people for a small fee. That this guy was making a living is a testament to the excitement of electricity flowing through human tissue.

Actually, current can be flowing through your skin without you even feeling it. Clamp a double-A battery between your thumb and finger and a small current will flow.

Feel anything? I didn't think so. We need to crank it up a bit to feel alive! That's where the transformer comes in. You've probably seen big ones perched on electrical poles and maybe seen small ones inside home appliances. Both of these applications are transforming a large voltage to a small one—"stepping down the voltage," in the technical lingo. Here we'll use a transformer in the other direction, stepping up the 1.5 volts of a common battery to a level that may make you yelp.

Gather stuff

- Battery
- Transformer (Any one will do, and big ones are no better or worse than small ones.)
- Four wires
- Tape
- Partner, if possible

Gather tools

- Wire strippers or knife
- Soldering iron, if you have one
- Multimeter, if you have one

TINKER

First you have to figure out your transformer. It will look something like this:

You'll find it has two sides, each one with at least two connection spots. Here is a diagram of what it looks like inside:

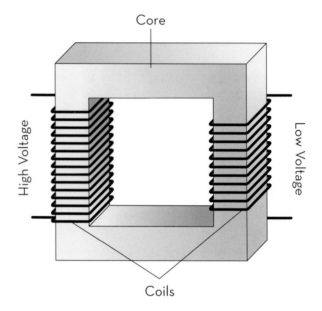

See that the two sides are not connected? They're complete wire coils with two connection points each. If your transformer has more than two connection points on one or both sides, its schematic diagram looks something like this:

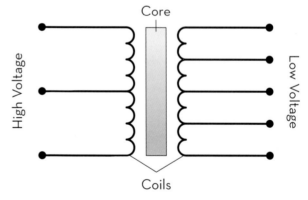

The coils are now represented by the bumpy lines and the core by the thin rectangle in the center. And in this schematic you can't tell by looking which side has more coils, but the voltage change depends on the number of coils.

Your transformer may be marked with numbers, maybe 220, 110, and 0 on one side and CT, 6, 9, 12, 18, and 24 on the other. CT stands for center, meaning the center of the coil, the middle branches on the schematic above. These numbers are a bit misleading, because what you get out depends on what you put in. So if you hook up 220 volts across the 220 and 0 points on one side, you'll get those other voltages on the other side, but if you don't you won't. We don't want anywhere near 220 volts, and we're only going to put in 1.5 volts, so the numbers written on the transformer don't describe what we'll see.

The main point is that the voltage is transformed from high to low or low to high, depending on how you hook it up. As I mentioned, normal small transformers like these are commonly connected to the wall outlet voltage and used to drop that voltage to a lower one. We'll hook ours up the other way to raise the voltage.

Find any two of the connections labeled with lower voltage, maybe 6 and 0 (or CT) and attach two wires. Solder them on if you want. Then connect two wires onto the other side, maybe at 220 and 110. You should now have four wires attached. The exact places are not so important; two wires on each side should do it.

Now you're ready to go. You can have your partner do this, but you can do it yourself as well. Use your finger and thumb to connect the two wires on the high voltage side. You can't just touch one; you've got to complete the circuit.

When that's complete, hook up the battery across the two wires on the low voltage side.

Is your heart yet aflutter?

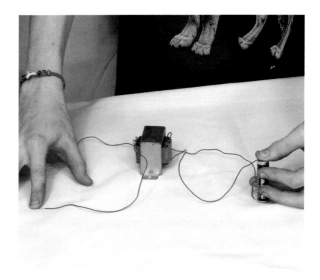

In this arrangement you've got one hand touching the higher voltage wires with two fingers. Thus, the electrical current is flowing down one finger, across the hand, and up the other finger.

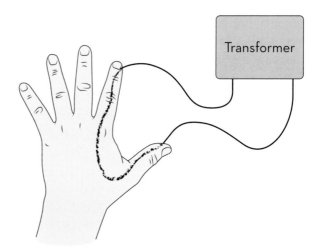

Another way to connect is by grabbing one wire with each hand. For this one you'll need a partner or clever switching mechanism to hook up the battery. The current now has to run all the way down one arm, across your chest, and up the other arm.

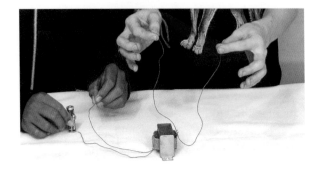

You may notice that your heart, not the least important organ of the body, is right in line with that circuit. More on that later.

You can also hold hands with someone else, or many other people, making a long, happy chain of fools—I mean scientists—all waiting to experience the thrill of electricity. (I've done this with more than 30 kids in a chain, some as young as 7, and got only giggles and joy.)

One final experiment to reference when I explain things in a moment: have your partner hook the battery up securely on one side and grab the two wires on the other side.

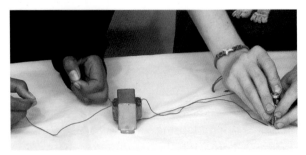

You will feel nothing; so far, so good. Now have your partner *disconnect* the battery.

Ouch. You'll get the same shock you did when the battery was connected. Now there's a stumper.

WHAT'S GOING ON?

The phenomenon at work here is electromagnetic induction. Any questions? No, but really, that $100 word stands at the culmination of electricity and magnetism. I introduced this section on electricity and magnetism with the transformer shock because it's so cool and memorable, and not at all because it's an easy place to begin explaining. The other activities in this chapter will explain a bit more about electrical circuits and magnetism.

Here we've fed a battery-powered current into a coil, which becomes an electromagnet. There is a second coil inside that transformer seated just beside the one getting the battery's current. When the first coil's magnetic field builds up, it causes a current to flow in the second coil; that's called *induction*, because it's an induced current. It requires both electricity and magnetism; thus it's electromagnetic induction. (Both electrical induction and magnetic induction also exist; feel free to have a go at finding out what those two are all about.)

It's critical to note two things about the transformer. First, the two coils are not electrically connected. You can prove this with a multimeter or by taking apart the transformer. Second, the coils are not the same size. If you take apart the transformer, you'll find the coil on the side labeled with the high voltages has more coils.

Now the key to it all: induction happens only when the magnetic field of the first coil is either building up or collapsing. That happens when the current is increasing or decreasing—that is, just as you connect it or just when you disconnect it. If the magnetic field is standing strong and steady—the circuit is complete and the current is flowing strong and steady—you get no induced current at all.

It's a bit mystical, eh? That's what the first scientists to note this effect must have thought.

There he was, Hans Christian Ørsted, a teacher doing a science demonstration lecture in 1820 Denmark with a table full of marvelous equipment, both electric and magnetic. Right in the middle of it all, he noticed a magnetic compass twitch when he completed a nearby circuit.

This is nothing to sneeze at. That finding—the linking of electricity and magnetism—was arguably the most significant discovery in modern physics and engineering. Within a year, Michael Faraday had constructed a working electrical motor using this concept, and to date the hours of mindless manual labor saved by electric motors must surely number in the gazillions.

That was mystical enough; an electrical current could create a force the same way chunk of magnetic iron can. But that wasn't yet the concept here! It was ten more years before Faraday discovered electromagnetic induction—the stunning fact that one current can set up a magnetic field that in turn induces another current.

This paved the way for transformers and voltage control. Check out this photo of Faraday's transformer here: way high tech, 1840.[1]

CREDIT: WIKIPEDIA COMMONS

Today, transformers are used all over the place to transform voltages. At power plants voltage is bumped up into the hundreds of thousands, which makes power transmission more efficient. It's also more deadly, so high towers keep the wires far out of reach. Before the electricity comes back down and into your house, it goes through another transformer that steps the voltage back down to around 120V in North America, or 240V in most of the world. This much voltage often does not result in death upon touching, especially if you're not wet. It also helps if you touch it with only one hand, and if you're wearing shoes.

Another safety tidbit: When professional electricians are looking into a problem or exploring a snarl of wires from unknown sources, you'll often see them with one hand behind their back. (Don't try this! Call one of them!) With this one precaution they can prevent electricity traveling right through their heart, as it does in our activity when you grab the two wires with your two hands. Instead, if they happen to touch some rogue electricity, it will attempt to travel through their entire body to the earth, which is a giant neutral sink for electrons. That's where the shoes come in; rubber soles increase the resistance and thus reduce the current. You may also notice the professionals stand on a wooden or plastic chair to further reduce any unforeseen current.

So fridges and ovens and toasters all use 120V in the United States, but things like computers and clocks and stereos don't need that much. These things generally have a transformer inside them, which takes the 120 volts down to 12 or 6 volts or just what the thing needs.

Getting back to our phenomena, we have one final concept to learn: DC and AC current. DC means direct current, and it's what our batteries give. Electrons come out of one side, flow directly through the circuit, and enter the other side. AC means alternating current, and this is what your wall sockets provide to your lights, fridge, and desktop computer. Alternating current is always going back and forth, changing from positive to negative voltage. A transformer will continuously pass this energy from one coil to the other, as the magnetic fields rise and fall continuously.

Transformers are almost exclusively used with alternating current. In this activity we're hacking the transformer—in essence, giving the current a single alternation (from off to on, or from on to off), and so you only ever get a single jolt. (Unless of course you have an evil mind and a jittery hand, and tappity-tap the wire on the tip of the battery—this will give your trusting partner multiple jolts, thus destroying the friendship for years to come.)

How much voltage are we generating here? I checked this on my oscilloscope, a handy device that plots an electrical signal on a graph of voltage versus time. I got around 40 volts maximum at the spike and about 50 milliseconds for the duration of the current. Many people with dry hands I've tested over the years (including myself) have a resistance of around 200,000 ohms, so with a useful equation called Ohm's law, Current = Voltage/Resistance, you find that the current is 0.2 milliamps. Danger comes only near 100 milliamps *through the heart*. This is a safe activity.

By the way, there is strength in numbers here. If you join a group of wild adventurers to make a ring holding hands, then the resistance of your circuit is more, so the current will be less for a given voltage. So enjoy safely shocking yourself and your friends. As Sia sang (screamed) over and over in her megahit, nearing a billion hits on YouTube as I write: I Love Cheap Thrills!

ENDNOTE

1. I'm a huge fan of Faraday. He self-studied his way up from an unschooled family into top levels of academia to make a long list of critical discoveries in diverse areas of science. Many snooty scientists of the day dismissed him due to his unimpressive pedigree, his lack of formal credentials, and his weak math. But James Clerk Maxwell, author of the comprehensive set of four equations linking electricity, magnetism, and light, saw that Faraday was way ahead of the crowd and used his results to assemble those equations.

Chapter 24

SEMICONDUCTOR CIRCUITS FROM THE TRASH

Use old holiday lights to tinker a diode circuit.

Y ou can get holiday lights for cheap, or even free from your neighbor's trash can—entire strings of lights that already have wires connected to them! This is an international phenomenon; we can get them here in Timor, too. They're everywhere at Christmas in this Catholic country, and when they stop working in a year or two, we get them for the asking!

What's more, many of them are now built with LEDs (light-emitting diodes), extraordinarily high-tech bits of solid-state physics that can be used to learn about semiconductors. LEDs also use only tiny trickles of electricity, so you can use smaller batteries and expect them to last for days.

Follow me through a few basic tinkerings we do here to learn about circuits, and I promise to show you a fascinating little gadget that can control the direction of a current.

Gather stuff

▶ String of holiday lights, preferably broken, plug chopped off, LED type. If you don't know if they are the LED type, you will find out in the first activity. Old-style, non-LED incandescent types can be used for some of the activities and to make interesting glowy projects.

Old style, non-LED type

LED type

- ▶ AA batteries, 6
- ▶ Aluminum foil
- ▶ Connection wire, around 28 gauge
- ▶ Electrical tape

Gather tools

- ▶ Wire stripper or knife
- ▶ Scissors

TINKER

First step: cut off the plug and throw it away.

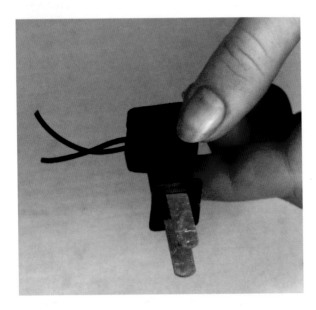

Usually I throw nothing away, but this is dangerous—you don't want anybody plugging the string of lights in after you modified them.

Now, figure out which wire the lights are strung along and how far apart they are, and then chop a few out midway between the lights. The other wires can be used to connect up other parts of the circuit.

Strip both ends of the wires from a few of the lights, maybe five. Wire strippers are pretty

useful, so I'd say you should get a pair, but if you don't have any, a knife or scissors will do the job.

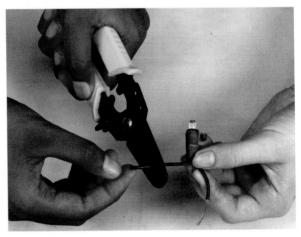

Connect one light to one of the AA batteries. If the wires are not long enough, cut another little piece, strip the ends, and connect it on.

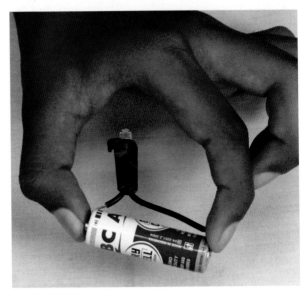

Did your light not glow? This is a very good indication that it's an LED! Common LEDs need more than 2 volts to light up, and your battery only gives 1.5 volts, as you can read on its side. If it did light up, it's probably an old-style incandescent light.

So add another battery, tail to head, to give you 1.5 + 1.5 = 3 volts. You can roll them in a piece of paper and tape it up to keep them together. You need to be able to see both ends.

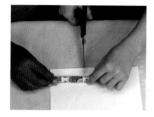

Now connect the light again and see if it glows. If it doesn't reach, add the other wire.

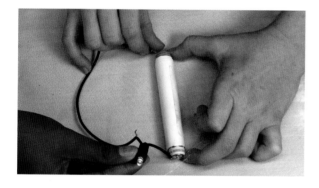

Ours lit up! If yours didn't and even if it did, *switch the wires around* and connect it again.

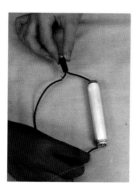

Interesting, eh? Darkness.

WHAT'S GOING ON?

That's what diodes are all about: the electricity goes one way and doesn't go the other. Diodes are the simplest of the semiconductors, and these materials are fairly new—that is to say, when I was young, they were super new, and when my parents were young, they hadn't been invented yet.

Usually when you hear of semiconductors it's in relation to transistors. Transistors are like little switches that can be used to make one circuit control another circuit. That's the beginnings of artificial intelligence, my friend, and when the first practical one was developed in 1947, it gave a huge kick to technology. Before transistors, you could still get one circuit to control another circuit, but it took a vacuum tube, which is a hot, energy-sucking glass bulb around the size of half of a sausage, and computers need hundreds of them to do the simplest thing. Today's computers have millions of transistors inside integrated circuits the size of your fingernail. So clearly, this was big stuff and continues to be.

A diode is like half a transistor and serves as a gatekeeper in a circuit. How do those incredible little buggers do that? Inside each diode there is a two-layer semiconductor wafer that is predisposed in one direction to reject one type of electricity—positive or negative—and let the other pass. You should definitely read up more on how semiconductors work, but for now I'll just repeat that what you've now witnessed in that tiny LED was not possible 75 years ago, so you should be impressed.

So the LED knows which way the electricity is going and only lets it through one way, at which time it glows. If you connected up your little holiday light and it lit up no matter which way it was connected, you've got an old-style incandescent light. You can't make the diode circuit that follows with these lights, but you can still make fun glowy things.

TINKER ON

Let's increase the voltage! Make a roll of four batteries. When they're head to tail like this, it's called series and the voltages add together. (Each battery is 1.5V, so how much is that total?[1]) Get two of your lights and hook them up to the batteries so they're right beside each other. Switch the wires around until they both light. Super bright, eh?

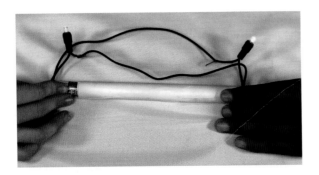

The batteries are in series but the lights are in parallel. The wires are sort of parallel, so it makes sense, eh? Now keep track of how the two LEDs are connected and hook them up in series, one after the other.

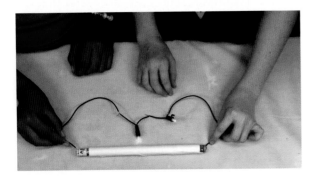

Which arrangement made the lights brighter? Our LEDs in parallel were much brighter than when they were in series.

This makes sense, right? The parallel lights are getting 4 × 1.5 = 6 volts now, because each

is connected directly across four batteries. The series lights are getting 6 / 2 = 3 volts now, because they're splitting the four batteries' 6 volts. Do they look about as bright as they were individually connected to two batteries?[2]

You may have noticed by now that sometimes it's a pain in the neck not knowing which way to hook up a battery to one of these LEDs. Fear not: I've got the solution. It's called the Full Wave Bridge Rectifier (FWBR). Build this and your problems are all behind you.

First you need to mark your LEDs so you know which side goes to the positive (bump) end of the battery and which side goes to the negative (flat) end. Test four of them and mark them with tape or a marker. We put little tabs of tape, but later we realized that the little hook on the side of each one showed which was negative. The tape was still better and easy to see.

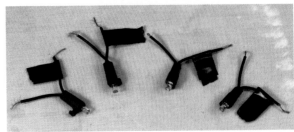

Now hook them up like this by twisting the stripped ends together. Trust me on this for now; I'll explain it later.

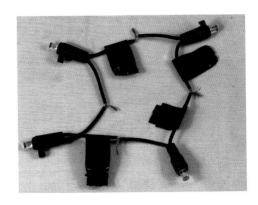

There is your Full Wave Bridge Rectifier. I'll describe it more fully in a moment.

Mark a fifth LED with some sort of special marking to differentiate it from the others, and hook it up across two opposite points, as shown in the photo here. You have to get those two connection points right on the FWBR, and you have to get that special LED right too. We put our special white tape on the negative side, just like all the black tapes on the other LEDs.

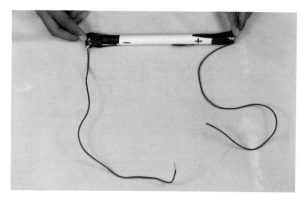

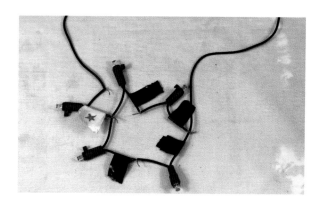

Now is the moment you've been waiting for. Hook up the batteries to the two other nodes on the FWBR *whichever way you want! And then switch the wires around and hook it up the other way!*

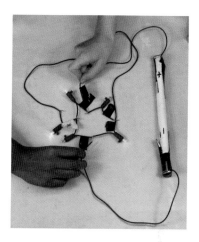

You'll need the long battery pack for the next part, or one even longer, and you'll want the connector wires stuck on it. We'll put little aluminum foil mittens on one stripped end of each wire to help make better contact with the battery. You want to fold the aluminum foil over the wires in such a way that the wires are bent inside the foil and don't pull out easily.

Stretch the tape tight to make the connections good. We put in six batteries and marked the positive and negative ends.

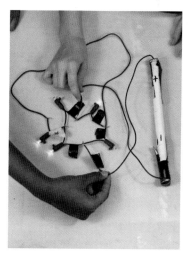

Your special LED should glow now, no matter which way you hook up the batteries. Also, two of the other LEDs in the FWBR should glow, showing you which way the electricity is traveling through it.

WHAT'S GOING ON?

Here are a few diagrams worth a bundle of words. First, the schematic diagram for the FWBR:

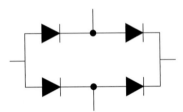

Those triangles with cross lines are the four diodes (ours light emitting!). The triangles are like arrows pointing the way the current can go through; the short lines crossing perpendicularly can be thought of as stopping current that tries to come in that way. Current is understood to travel from positive to negative.[3]

You have four places to hook up to now on the FWBR, called nodes, the four sides of the square. You can see that the two nodes on the sides enter and encounter only one type of LED side: right side stopping current, left side receiving it. The top and bottom nodes enter and have both sides to choose from: go left and get stopped, but go right and get received.

That's just the situation in the one we built. Here's a diagram of the circuit with that special extra LED drawn off to the left, showing the electricity flowing when you hook it up one way. Check all three diodes that are getting current and making light; the electricity is passing in the direction of the arrow each time, from the battery's positive terminal to its negative terminal.

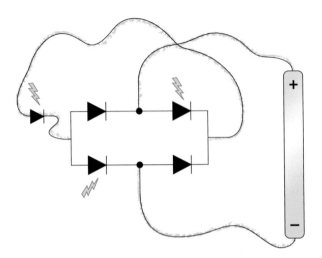

Now here is the diagram showing the electricity flowing when you hook it up the other way. Again, current is going through right with the arrows, + to –, not passing through the other LEDs the wrong way.

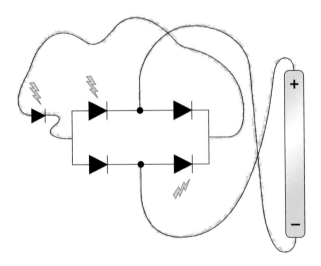

So no matter which way you hook the batteries up, the current finds a way to flow through. Pretty ingenious, eh?

The FWBR is used all over the place, and it's not just to make life easy when hooking up batteries. What it's really good at is turning alternating current electricity to direct current electricity. DC electricity is the kind we've been working with in this activity. It comes from batteries or power supplies and involves a current traveling from positive to negative, always in the same direction.

AC electricity is what is coming out of the wall plugs and light sockets of your house. There are no positive and negative wires; rather, the current changes direction every 1/120th of a second. Sounds strange, but you can transfer energy just as well this way, and it turns out AC is much more efficient than using DC over long distances.

But many small devices, including computers, phones, and stereos, use DC, so before the AC electricity from your wall socket enters your device, it needs to be converted into DC. Full Wave Bridge Rectifier to the rescue.

Just as we saw, it doesn't matter which way we feed electricity into the two input points; at the output points, the LED got just what it needed. We can actually label those output points as positive and negative; they will never change.

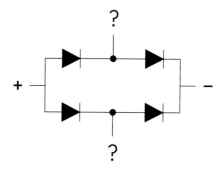

The other two nodes are marked with question marks, because it doesn't matter which side gets connected there. In an AC-to-DC circuit, the AC wires get connected to these to nodes. As the AC direction flips back and forth, back and forth 120 times per second, the FWBR takes it all in stride and feeds positive to the positive point, negative to the negative, creating DC at the side nodes. Slick, huh? I'm impressed.

ENDNOTES

1. You could use a rectangular 9V battery here. We have a hard time finding them here, and they don't last as long as four or six AAs, so we just make this pack.

2. You may have thought of this already, but the natural way to know how much electricity your LED can take is...to give it more than it can take! If you've got plenty of batteries, and plenty of these little LED dandies, go ahead and sacrifice one or two. Stack up the batteries, adding one by one, until your LED is no longer glowing merrily with the electricity but has reverted instead to a state of permanent, dark indifference. Then count the voltage and you'll know what the limit is. I guess that's beyond the limit, on second thought.

3. It turns out electrons make current and they travel from negative to positive. It was a historical technical glitch we've got to live with now. So just be clear when you speak of directions in a circuit: are you talking about the "classical current" or the actual electrons?

Chapter 25

TAKING APART WATER

Bust apart the H_2O molecule with a small current.

When we boil water, we bump individual molecules from a liquid to gaseous state. We do that by increasing the energy, in the form of heat, which increases the speed of the molecules. They bounce up out of the liquid, away from each other, and are no longer confined to the pot. Water molecules in the gaseous state are still affected by gravity, drawn to the earth like the rest of the gas molecules in the atmosphere, but now they have enough energy to bounce off each other and whatever surfaces they come upon, flying freely above the solids and liquids below.

That's called a "physical change" or a "state change," because you're changing the state of the water from liquid to gas. If instead you yank apart the atoms that are bonded together to form the water molecules, it's called a "chemical change." To do this you also need energy, and you can provide it with a steady stream of electrons. It's called electrolysis, and it's easy to do this

with a few batteries. On the Internet you can see sleek, professional laboratory setups. In Timor we do it in a scrap bottle; let me show you how.

Gather stuff

▶ Cups

▶ Water

▶ Salt

▶ Other things to dissolve in the water: soap, vinegar, alcohol, lime juice, sugar, bleach, Epsom salts, etc.

▶ Strips for electrodes: aluminum foil, galvanized nails, coins, pieces of a soup can, pencil leads (either from mechanical pencils or whittled out of normal wooden ones), etc.

▶ Electrical wire; around 28 gauge works well.

▶ Battery pack: 6 or more Ds works well, though a big ol' 12 volt or two 6 volts would

247

work even better. You can also use a battery charger or power supply.

- ▶ Plastic bottle with lid
- ▶ Duct tape or packing tape
- ▶ Matches
- ▶ Distilled water, if you have some
- ▶ Stainless steel bolts, if you can find some
- ▶ Small cups or ice-pop tubes
- ▶ Low, wide container
- ▶ Twist ties or other thin wire

Gather tools

- ▶ Knife
- ▶ Scissors
- ▶ Wire strippers, if you have some

TINKER

Stir a couple of spoons of salt into a cup of water. Hook up some wires to your battery pack, strip the wires' ends nice and long, maybe 5 cm, and stick these stripped ends into the cup of saltwater. Look very closely to see what happens.

Do you see tiny bubbles forming around one or both of the wires?

There's your experiment: electrolysis in action. End of story.

No, no, just kidding—not at all the end, but that's the basis of it anyway. When you feed electrons (electricity) into some water, you get a chemical reaction, and at least one product of that reaction is gaseous, making bubbles. Actually, you can call it an electrochemical reaction, because electricity is needed.[1]

> **SAFETY NOTE** Chlorine gas may be produced in small quantities in this activity, and it's not good for humans to breathe, so don't be inhaling too much near the top of your cups, and stay in a well-ventilated area.

Now that you've seen the bubbles, note which side they come from: positive, negative, or both. Also look closely to see if there is anything else happening there aside from the bubbles. You may be able to see crud of a strange color forming or flaking off the wires. That's another sign that a chemical reaction is happening.

You can do several things to get more information on what's happening in the cup. First let's try changing the water. Get a clean, dry cup and fill it with distilled water if you have it, tap water if you don't, and try the experiment again.

What happened? In theory, the more pure your water is, the less current will flow and the fewer bubbles you'll see. More on that later.

Now let's dump some other stuff in the water. We're trying soap and vinegar in the photos here, but you could put in whatever else you want, like the Gather Stuff list. Get several cups and compare the ones you've got. Stick the wires in one by one and see if you can detect a difference between them: more or fewer bubbles, different kinds of crud forming, or whatever.

Now try changing the electrodes. Electrodes are the things the electrons jump off of into your experiment, so up to now you've had copper wire as your electrodes. Did you notice one or both of them seemed to change color or form crud?

Try a couple of other materials from the list. We're trying galvanized screws and aluminum foil in the photos here. If you've got stainless steel bolts—available one by one at the local hardware store and rather expensive—try them too. Try these electrodes in each of the different solutions and see what happens.

You may want to stop and record some of this; we've got a lot of combinations now, and a lot of observations on each one. A lab notebook is always a useful thing, especially next week when all these details will be fading from your memory.

The next step will be to try to figure out more about these gases. That means we need to try and collect some of them. So as you tinker with these different combinations, figure out which stuff in the water together with which electrode material gives the most bubbles. We're going to use aluminum in saltwater because it gives us many bubbles from both sides.

You should find at least one combination that gives plenty of bubbles from both the positive and negative electrodes. If you haven't, or if your bubbles are generally puny and unimpressive, you'll need to up your voltage and/or your current. Bigger batteries or a battery charger will do the trick. Don't go much over 20 volts to avoid any possibility of a dangerous shock.

First let's build the vessel. Make two slits near the top of the water bottle.

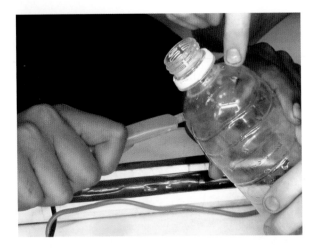

Jam the electrodes into the slits and twist the wires onto the electrodes. Tape up the holes tightly so they are more or less airtight.

Now fill the bottle with the solution you found gave the most bubbles. Stop a few centimeters before the slits so that the water doesn't leak out. There should be a small space of air up near the neck of the bottle. Then put the lid on, hook it up, and watch it bubble. We've got the battery charger hooked up here: 12 volts, nice and steady.

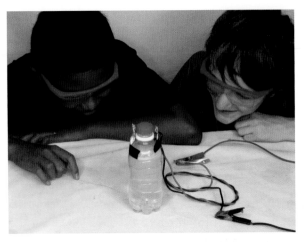

After five minutes of bubbling, you can imagine that these bubbles have all gone up into the space above the water. We're trying to find out more about these gases, and one common way to find out more about an unknown gas is by exposing it to fire.

SAFETY NOTE Before you burn anything, get an adult, put on your safety glasses, get your fire extinguisher ready, and keep your face back.

Light a match, hold it close to the lid, open the lid a tiny bit, and move the flame into the mouth of the bottle.

Did it whoof? If it didn't, try again. Make sure a lot of bubbles are rising, and wait twice as long this time. Also, make sure there is not too much air in the top of the bottle. Too much air will dilute the gases we're trying to understand, so it's best to start with just a bit.

What do you think the whoof tells us? Well, most basically, it tells us that the bubbles coming up are different from the air around us, which doesn't generally whoof when exposed to flame. It tells us that one or more of the gases produced are ready to enter into a combustion reaction when given enough heat.

But before we analyze that any further, we can try to separate the bubbles coming from the two sides.

This is not so easy in a plastic bottle, so we'll move to a small, low container. The trick to isolating the bubbles is to upturn some small vessel in the water so that it is filled with the water solution and whatever bubbles come up are the only gas in the vessel. You want the vessels to be transparent so you can see how you're doing. Test tubes are often used, but you can use little cups too. We take advantage of the slender vessels those ultra-sweet ice pops come in.

The logistical trick is to fasten the vessels down so they don't tip over or float off. We poked holes in the side of our container so that we can tie the ice pop tubes to the sides with twist ties! We melted the holes with a nail heated with a candle, but you could use a drill.

> **SAFETY NOTE** Get an adult to help you if you use the melting process.

Prepare the wires and electrodes to be stuck up well inside the vessels so that the bubbles don't escape. Fill the container with your solution; we're using saltwater again. Put the vessels under the water so that they fill up and no bubbles are left at the top, and then upturn them and fasten them to the sides.

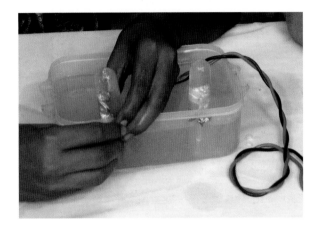

Now don your safety glasses, turn on the juice, and watch the bubbles brew.

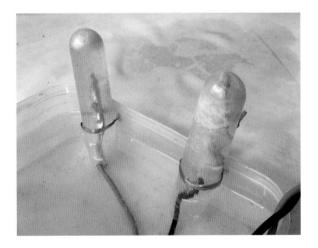

Wait until one or both of the vessels are full of gas. Before you haul them out to test, compare the volume of the gas in the two vessels: is it 2 to 1, like the H_2O formula suggests? Don't be surprised if it's not—a number of complications arise to complicate this complicated experiment.

Now it's time to test the gases with the flame again. It's a bit tricky, but you can handle it. Keep your safety glasses on. Unhook the electricity and carefully pull out the electrodes without losing any of the gas. Light a candle. Untie one of the vessels. Slip your finger underwater and cover the end of it.

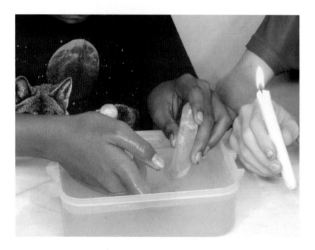

Take it out of the water, tip it up, and get ready to light whatever is inside it. Open the end of the vessel very near the candle and see what happens.

Do the same to the other vessel and think about what you observe.

Did one or both whoof? Did one of them make the flame flare up glowingly for a moment? Or did they do nothing at all?

Of course, things this complex and funky rarely work the first time, so be ready to set it all up again, tweak your arrangement for ease of testing when it's done, turn up the electricity a bit, and wait for more vessels-full of gas to test.

WHAT'S GOING ON?

The textbooks say that water is composed of hydrogen gas, which is highly flammable, and oxygen gas, the base of combustion reactions. When we saw bubbles coming out of water upon the introduction of electricity, that was our first indication that the textbooks may be right, but you must be careful about what you actually conclude from this exquisite dance of electrochemistry.

We saw that the mixture of the bubbles forming at both the electrodes, and the air above the water, was pleasantly explosive. When we managed to separate the bubbles coming from positive and negative sides, we sometimes got that one side was burnable—whoof—and the other increased the level of a flame. That also matches theory: oxygen would increase a reaction and hydrogen burns. Put them together and they'll really burn, even when mixed with some air, which also has oxygen in it.

If one of your separated vessels happened to contain twice the gas of the other one before you tested it, congratulations; there is twice as much hydrogen as oxygen in water. Now be sure it was the right one—hydrogen should be produced at the negative electrode, oxygen at the positive electrode. So the vessel with the negative electrode—the one connected to the negative side of your battery pack—should, if everything worked just right, contain twice as much gas as the vessel with the positive electrode. And the gas from the negative side should whoof while the gas from the positive side should cause the match flame to flare. How did you do?

If you got any of these results, that's quite significant, to be sure, and you can be happy that you've come so far in gaining evidence for what's in the textbooks, and with just the junk we've got lying around on the table. We've done this experiment many times, and many times we've ended up with no conclusive results at all. But other times we have, so if yours doesn't work the first time, please believe it's possible and give it another shot.

But even if you got good gas production in both vessels, and one whoofed and one flared, to say you've proved water is made from two parts hydrogen and one part oxygen is a little bit of a fib. We'd need a lot bigger lab to do that.

Why do I say that? Well, a whole number of gases burn when exposed to flame, and there are a lot of other reactions running in the setup you've just presided over. Your electrodes probably changed color and/or disintegrated into the solution, and the solution probably changed color and developed some muck in the bottom, eh? All that means your reactions were more complicated than the straight production of H and O.

Let's look again at what's happening in the solution. Ions (see the sidebar) are necessary if you want the water to conduct electricity. Electricity is free electrons moving from atom to atom, pushed by a voltage. In a wire, that happens to electrons in copper atoms. Metals in general have an upper level of free electrons ready to move when pushed. But a water molecule doesn't, and that's why pure water—no ions—is an insulator. Problem is, pure water is nearly impossible to find or make. It's almost always got many minerals and salts and other odds and ends dissolved in it, so as a rule, water is a fine conductor because of all the ions.

ELECTROLYSIS EQUATIONS

Here's a look at the equations, ultra-simplified. When you have a cup of water, it's full of water molecules, but also of H^+ and OH^- ions. An ion is a sort of half-molecule that exists in solution. Ions are charged, meaning they don't have an even number of electrons for the protons in the nuclei. The ones with extra electrons are negatively charged—here that's the OH^- ion, called hydroxide—and the ones short on electrons are positively charged—here that's the H^+ ion, just called a hydrogen ion.

When the electrons coming from the negative electrode hit the solution, ions nearby there are the first to react. The hydrogen ion picks up the first electron it can find and turns into a hydrogen atom, which binds to the first other hydrogen atom it can find and becomes the H_2 molecule—pure, stable hydrogen. Here's the equation:

$$2H^+ + 2e^- = H_2$$

The OH^- ions next to the positive electrode are itching to get rid of an extra electron. When you hook up the electricity, the positive electrode can be imagined like it's sucking electrons out of the solution. Those electrons are present in the negative OH^- ion. When OH^- kicks out an electron, it is ready to recombine to form a neutral oxygen molecule and a water molecule. Actually, you need that to happen four times to get two oxygen atoms to join together, because like hydrogen, oxygen is most stable as a two-atom molecule, O_2. Here's the balanced equation:

$$4OH^- = O_2 + 4e^- + 2H_2O$$

You see the little e's in those equations? What they signify is electricity! One is on the right and the other on the left, meaning these reactions go great together; one needs e's and the other gives e's. That's the electro part of electrochemistry!

Here, we've dumped in salt or vinegar or whatever, which introduced many more ions into the solution, and those ions helped carry the current. You can imagine the electrons hopping freely from one ion to the next. But the ions also muddied the reaction possibilities. Instead of just those nice clean equations in the sidebar, we also had reactions happening with the ions of salt—Na^+ and Cl^-—and the materials of our electrodes—aluminum, zinc, copper and iron. Some of these other reactions might steal the gas you're trying to generate and use it to form an ugly precipitate, the crud that forms and falls to the bottom.

One way to avoid unwanted reactions is to use something that doesn't easily react. Platinum and gold are possibilities, but please don't screw up your mom's gold earrings to improve this experiment. Stainless steel is "stainless" precisely because it doesn't react easily with other things. If you try your experiment with stainless steel bolts or perhaps spoons (again, ask your folks first!), it should produce little to no precipitate.

Which reactions are actually happening? That's the million-dollar question in chemistry. It's also got a standard answer: they're all happening, all the time, but not at the same level. Chemists often speak in terms of which reaction is "favorable" or "likely." They figure this by looking at the electrical energy needed to make each one happen. This is called the *reaction potential*. Stick that term in your pocket for future use.

You can see how all these possibilities make it hard to the point of impossible to predict what you've got bubbling out of your solution. But, once again, we can tentatively conclude that if it whoofs, there was probably some hydrogen in there. And if it made a fire burn brighter, or a glowing stick flame up, you likely had some oxygen. Well done, if you saw that.

Please don't think about scaling this up. Hydrogen explosions are famous in history, and not the kind of famous you want to be: dead famous. Be content with a nice little whoof and a new perspective on the substance you drink every day.

ENDNOTES

1. By the way, inside your batteries other electrochemical reactions are happening, some that need electrons and others that produce extra electrons. That's how batteries work, and you can make one with two kinds of metal and any acid, base, or salt. Try it out!

ABOUT TIMOR-LESTE

First, here are some key dates in Timor-Leste's history:

42,000 years ago—Oldest archeological find in Timor with human remains.

1225—Notes of Chinese official Zhao Rukuo mention Timor as source of quality sandalwood.

1600s—Portuguese invade Timor, set up trading post, and begin exploiting sandalwood resources.

1749—Timor split following battles between Portuguese and Dutch. Portuguese take the eastern half along with a small enclave in the west.

1942—Japanese invade Timor, and clash with Australian troops who are fighting there to avoid a Japanese invasion of Australia. Some 60,000 Timorese are killed under Japanese rule, which ends in 1945.

1974—Coup in Lisbon leads to new Portuguese government that begins policy of decolonization.

1975—Timorese Fretilin party unilaterally declares an independent Timor-Leste. Indonesian troops invade on December 7. More than 150,000 people—a quarter of the population—killed by fighting, famine, and disease that follow the invasion and during 25-year Indonesian occupation.

1975—First of eight UN General Assembly resolutions about Timor-Leste condemns Indonesia's invasion and calls for troop withdrawal; the United States opposed or abstained from each resolution.

1991—Santa Cruz Massacre, in which more than 270 peaceful demonstrators were killed by Indonesian military forces; Briton Max Stahl captures massacre on film.

1996—José Ramos-Horta and Bishop Carlos Filipe Ximenes Belo receive the Nobel Peace Prize for their work to resolve the ongoing conflict.

1999—After leadership changes in Indonesia, UN facilitates accord allowing a referendum on independence in East Timor on August 30.

1999, September—After 78% of voters opt for independence, Indonesian military and their militia increase campaign of terror: 80% of infrastructure destroyed, more than 1,000 people killed, and upward of 300,000 forcibly dislocated before Indonesian forces leave East Timor.

1999, October—Peace and freedom in Timor-Leste as UN takes over transitional administration and helps to prepare the nation for independence.

2002—Timor-Leste becomes first independent nation of the new millennium.

Most people to whom I've told the story of Timor-Leste's history are surprised to hear that the bloody acts of Indonesia had solid U.S. support. In fact, some 80% of the weapons employed were U.S.-made. U.S. President Ford and Secretary of State Kissinger were visiting Jakarta the day before the 1975 invasion, and upon hearing Indonesia's plans, asked only that they be allowed to leave Indonesia before hostilities began. U.S. support for the Indonesian military and their ongoing abuse of the Timorese people, which steadily increased for the next two decades, arose partly from the ongoing fear of the spread of Communism in the region and partially from geopolitical power wrangling. Indonesia is the fourth most populous nation and the largest Muslim nation, so as a Jakarta diplomat pointed out, often in U.S. foreign policy, "Indonesia matters, East Timor doesn't" (see www.mediamonitors.net/indonesia-east-timor-and-the-western-powers/).

During the 1990s I did solidarity work for Timor-Leste with my partner Pamela and the East Timor Action Network (ETAN) toward the goal of getting the United States to change its foreign policy and stop supporting the illegal Indonesian occupation. The story of ETAN stands as a testament to the effectiveness of international solidarity toward a just cause. Starting with a handful of concerned activists, mostly academics, ETAN grew to a membership of 11,000 by the late '90s. Success came in late 1997 when both houses of Congress voted unanimously in favor of bills that would end weapons shipments to Indonesia.

Pamela worked full-time in 1999 to help prepare for and monitor the UN referendum on independence in Timor-Leste. The photo below shows the population waiting to vote in the city of Suai. In a remarkable demonstration of democracy, 99% of registered voters turned out in spite of massive intimidation.

The violence following the vote was the agonizing end to the Indonesian era. UN forces restored order, but the place was a shambles.

PHOTO CREDIT: DIANE FARSETTA

Since late 1999, Timor-Leste has been working to develop as a modern, democratic, and sovereign nation. The Timorese continue to grapple with all the tricky questions of development priorities and nation building while enjoying a freedom that came at a terrible price. More up-to-date information can be found on the website for the Timorese organization called *La'o Hamutuk* (Walking Together): www.laohamutuk.org.

I documented our experiences here in Timor-Leste in the early years of independence in newsletters for the Institute of Current World Affairs, found online here: www.icwa.org/curt-gabrielson-newsletters/.

Most Timorese are subsistence farmers and don't really make money; they make food, and maybe sell a bit of it to buy a few consumer items like cooking oil and eyeglasses. But most folks in the world live more or less like this. Almost a billion of the world's people live on less than $2 a day, and the majority live on less than $6 a day. The average Timorese made around $4 a day last year, whereas I made around $160 per day here. That 40-fold ratio makes for a world of difference between my life and theirs.

To me, it means my life—in a global sense—is weird and highly privileged, whereas the average Timorese life is *normal*. It's the way things are for most folks around the world right now. So if you're reading this from somewhere like my home culture of middle-class United States, there's no use feeling guilty or sad, but please do be aware of this.

Actually there are plenty of things you can do to help move toward a better global income distribution and a more just use of the world's resources. Maybe start by checking the Facebook site for Permatil, Timor-Leste's leading grassroots organic agriculture organization, at "Permakultura Timor Leste."I know and trust these folks. We're impressed with their work and support their continued sustainable agriculture projects in Timor-Leste's schools and communities.

CONNECTIONS TO THE NEW GENERATION SCIENCE STANDARDS

When you read the NGSS, it's like a breath of fresh air. They are steeped in the fundamental tenets of science and have certainly shifted the emphasis away from straight information transfer. They absolutely give space and freedom to do science education right from a variety of angles, and to accompany students in their own authentic search for meaning and understanding within the broad glory of the universe, with your classroom as primary portal. What I see and hear happening in far too many schools, though, is not that.

It seems an unavoidable reality that sets of standards are overloaded. It just must be so friggin' tempting for the authors to put every morsel of good stuff in there, with the result that even if students ignored the rest of the disciplines and learned just science each year, there still would likely not be enough space to really teach these standards in completeness.

So in many schools today teachers are struggling to "get through" the NGSS. Ouch. Here in Timor-Leste the situation is the same. They call it "chasing the curriculum." And when you're just getting through something, it can't be made meaningful. What's more, when time is short, the pivotal and gritty process of construction of meaning through personal investigation is the first to be chucked out the window in favor of a list of important-sounding factoids to memorize.

I laid out a strategy for dealing with standards in the first *Tinkering* book. I suggest that teachers plan first to teach great hands-on activities that they know to be popular and effective and locally relevant, and then bend over backward to make the links to the standards, at whatever grade level. Then tell your boss you're "teaching to the standards"! In the following table, I've given possible links to the three dimensions of the NGSS; there may be others. Several activities integrate bio, chem, and physics all at once—can't go wrong. The math section is a bit weak on connections; check your math standards!

	NGSS PRACTICES								NGSS CROSSCUTTING CONCEPTS							DISCIPLINARY CORE IDEAS
	1. Asking questions (for science) and defining problems (for engineering)	2. Developing and using models	3. Planning and carrying out investigations	4. Analyzing and interpreting data	5. Using mathematics and computational thinking	6. Constructing explanations (for science) and designing solutions (for engineering)	7. Engaging in argument from evidence	8. Obtaining, evaluating, and communicating information	1. Patterns	2. Cause and effect: Mechanism and explanation	3. Scale, proportion, and quantity	4. Systems and system models	5. Energy and matter: flows, cycles, and conservation	6. Structure and function	7. Stability and change	
1. Coconut Oil	x		x			x		x		x	x	x		x		PS1.A, PS2.B, PS2.C, LS1.A
2. Remixing Air	x		x		x	x	x	x	x	x	x	x	x	x	x	PS1.A, PS1.B, PS2.A, PS2.B, PS2.C
3. Biogas	x		x			x	x	x		x	x	x	x	x	x	PS1.A, PS1.B, PS2.B, PS2.C, PS3.A, PS3.B, PS3.D, LS1.A, LS2.B
4. Fermentation and Distillation	x		x		x	x	x	x		x	x	x	x	x	x	PS1.A, PS1.B, PS2.B, PS2.C, LS1.A, LS1.B, LS2.A, LS2.B
5. Basketry in Timor-Leste					x			x	x		x			x		PS1.A, PS2.A
6. Baskets Full of Math	x			x	x	x		x	x		x					PS1.A
7. The Rhombus Weave					x					x				x		PS1.A
8. *Kohe*					x					x				x		PS1.A, ESS3.C
9. Stick Solids	x		x	x	x	x		x		x	x			x	x	PS1.A
10. Food Calendars	x		x		x	x	x	x	x			x	x		x	LS1.A, LS2.A, LS2.C, LS4.D, ESS3.A
11. Tinkering with Plants	x		x				x	x	x	x				x	x	PS1.A, PS2.B, LS1.A, LS1.B, ESS3.C
12. Finger Model and Bird Foot Dissection	x	x	x			x	x	x		x		x		x		PS2.A, LS1.A, LS1.A, LS1.B, LS4.B, LS4.C

	NGSS PRACTICES								NGSS CROSSCUTTING CONCEPTS							DISCIPLINARY CORE IDEAS
	1. Asking questions (for science) and defining problems (for engineering)	2. Developing and using models	3. Planning and carrying out investigations	4. Analyzing and interpreting data	5. Using mathematics and computational thinking	6. Constructing explanations (for science) and designing solutions (for engineering)	7. Engaging in argument from evidence	8. Obtaining, evaluating, and communicating information	1. Patterns	2. Cause and effect: Mechanism and explanation.	3. Scale, proportion, and quantity	4. Systems and system models	5. Energy and matter: flows, cycles, and conservation	6. Structure and function	7. Stability and change	
13. Heart Model and Heart Dissection	x	x	x			x	x	x		x		x		x		PS2.A, LS1.A, LS1.B, LS4.B, LS4.C
14. Sun Path Model	x	x	x	x	x	x	x	x	x	x	x	x	x	x	x	PS2.A, ESS1.B, ESS2.D
15. Global Warming Models	x	x	x	x	x	x	x	x	x	x	x	x	x	x	x	PS2.B, PS2.C, PS3.B, PS4.B, ESS2.D
16. Rock Cycle	x	x				x	x	x	x	x	x	x	x	x	x	PS1.A, PS2.A, PS2.B, PS2.C, ESS1.C, ESS2.A
17. Plate Tectonics	x	x	x			x	x	x	x	x	x	x	x	x	x	PS1.A, PS2.A, PS2.B, PS2.C, ESS1.C, ESS2.B
18. Flip-flop Air Gun	x		x		x	x	x	x		x	x		x	x		PS2.A, PS2.B, PS2.C, PS3.A, PS3.B, PS3.C, ETS1.A, ETS1.B, ETS1.C
19. Newton's Slingshot	x		x	x	x		x	x		x	x		x			PS2.A, PS2.B, PS2.C, PS3.A, PS3.B, PS3.C
20. Floating and Sinking	x		x	x	x	x	x	x		x	x				x	PS1.A, PS2.A, PS2.B, PS2.C
21. Lakadou	x		x		x	x			x	x	x		x	x	x	PS1.A, PS2.A, PS2.B, PS2.C, PS3.C, PS4.A, ETS1.A, ETS1.B, ETS1.C

	NGSS PRACTICES								NGSS CROSSCUTTING CONCEPTS							DISCIPLINARY CORE IDEAS
	1. Asking questions (for science) and defining problems (for engineering)	2. Developing and using models	3. Planning and carrying out investigations	4. Analyzing and interpreting data	5. Using mathematics and computational thinking	6. Constructing explanations (for science) and designing solutions (for engineering)	7. Engaging in argument from evidence	8. Obtaining, evaluating, and communicating information	1. Patterns	2. Cause and effect: Mechanism and explanation.	3. Scale, proportion, and quantity	4. Systems and system models	5. Energy and matter: flows, cycles, and conservation	6. Structure and function	7. Stability and change	
22. Tube Music	x		x		x	x			x	x	x		x	x	x	PS1.A, PS2.A, PS2.B, PS2.C, PS3.C, PS4.A, ETS1.A, ETS1.B, ETS1.C
23. Electrifying Induction	x		x		x	x	x			x	x	x	x	x	x	PS2.C, PS3.A, PS4.C
24. Semiconductor Circuits from the Trash	x		x	x	x	x	x	x	x	x	x	x	x	x	x	PS1.A, PS2.C, PS4.C, ETS1.A, ETS1.B, ETS1.C
25. Taking Apart Water	x		x	x	x	x	x	x		x	x		x	x	x	PS1.A, PS1.B, PS2.B, PS2.C, PS3.A, PS3.D

INDEX